The Cognitive Basis of Aesthetics

This book seeks to fill a void in contemporary aesthetics scholarship by considering the cognitive features that make the aesthetic and artistic worthy of philosophical study. Aesthetic cognition centers on the idea of art as a unique sensuous or imaginatively-intended form of meaning that transforms our self-understanding and comprehension of other people. This idea was central to the rise and consolidation of aesthetics as a discipline, but has been largely abandoned by analytical philosophy, which instead tends to focus its attention on the 'non-exhibited' properties of artwork or issues concerning semantic and syntactic structure. *The Cognitive Basis of Aesthetics* innovatively seeks to correct the marginalization of aesthetics in analytical philosophy by reinterpreting aesthetic cognition through an integration of Ernst Cassirer's philosophy of symbolic forms with Paul Crowther's theory of imagination and philosophy of art. This integration has three important outcomes: 1) it explains why the aesthetic and artistic constitute a unique form of knowledge; 2) it shows the role this plays in the formation of aesthetics as a discipline; and 3) it describes why aesthetic cognition is so deeply engaging. This book's unique theoretical approach engages with important works of visual, conceptual, and digital art, as well as literature, music, and theatre.

Elena Fell received her diploma in philosophy from St. Petersburg State University and her doctorate in philosophy from the University of Central Lancashire. She is Assistant Professor at Tomsk Polytechnic University and is the author of *Duration, Temporality, Self: Prospects for the Future of Bergsonism* (Peter Lang, 2012).

Ioanna Kopsiafti obtained her master's degree in art history at the University of St. Andrews and her doctorate in art history and theory at the University of Oxford. She is Assistant Professor at Hellenic American University in Athens, Greece, and is also an Adjunct Professor at the Rome campus of Loyola University Chicago. Elena and Ioanna are co-authors of *Thinking Space, Advancing Art: Cassirer and Crowther* (Cambridge Scholars Publishing, 2015).

Routledge Studies in Contemporary Philosophy

For a full list of titles in this series, please visit www.routledge.com

72 **Experiential Learning in Philosophy**
Edited by Julinna Oxley and Ramona Ilea

73 **On the Genealogy of Color**
A Case Study in Historicized Conceptual Analysis
Zed Adams

74 **Reification and the Aesthetics of Music**
Jonathan Lewis

75 **Intellectual Virtues and Education**
Essays in Applied Virtue Epistemology
Edited by Jason Baehr

76 **Embodied Emotions**
A Naturalist Approach to a Normative Phenomenon
Rebekka Hufendiek

77 **Normativity and Naturalism in the Philosophy of the Social Sciences**
Edited by Mark Risjord

78 **The Concept of Violence**
Mark Vorobej

79 **A Social Theory of Freedom**
Mariam Thalos

80 **The Cognitive Basis of Aesthetics**
Cassirer, Crowther, and the Future
Elena Fell and Ioanna Kopsiafti

The Cognitive Basis of Aesthetics

Cassirer, Crowther, and the Future

Elena Fell and Ioanna Kopsiafti

Routledge
Taylor & Francis Group
LONDON AND NEW YORK

First published 2016
by Routledge

2 Park Square, Milton Park, Abingdon, Oxfordshire OX14 4RN
52 Vanderbilt Avenue, New York, NY 10017

Routledge is an imprint of the Taylor & Francis Group, an informa business

First issued in paperback 2019

Library of Congress Cataloging-in-Publication Data
Names: Fell, Elena, 1975– author.
Title: The cognitive basis of aesthetics : Cassirer, Crowther, and the future / by Elena Fell and Ioanna Kopsiafti.
Description: 1 [edition]. | New York : Routledge, 2016. | Series: Routledge studies in contemporary philosophy ; 80 | Includes bibliographical references and index.
Identifiers: LCCN 2016003266 | ISBN 9781138652149 (alk. paper)
Subjects: LCSH: Aesthetics. | Symbolism. | Art—Philosophy. | Cassirer, Ernst, 1874–1945. | Crowther, Paul.
Classification: LCC BH39 .F454 2016 | DDC 111/.85—dc23
LC record available at http://lccn.loc.gov/2016003266

ISBN: 978-1-138-65214-9 (hbk)
ISBN: 978-0-367-87781-1 (pbk)

Typeset in Sabon
by Apex CoVantage, LLC

This research was funded by RHSF 15-03-00812a.

Contents

Figures

Abbreviated Titles of Works by Cassirer and Crowther Cited in the Notes

Ernst Cassirer

PSF—Philosophy of Symbolic Forms
Vol.1, *Language*, trans. Ralph Manheim, Yale University Press, New Haven and London, 1953
Vol.2, *Mythical Thought*, trans. Ralph Manheim, Yale University Press, New Haven and London, 1955
Vol.3, *The Phenomenology of Knowledge*, trans. Ralph Manheim, Yale University Press, New Haven and London, 1957
Vol.4, *The Metaphysics of Symbolic Forms, trans. J.M.Krois and D.P.Verene, Yale University Press, New Haven and London, 1996*
LCS—The Logic of the Cultural Sciences, trans. S.G.Lofts, Yale University Press, London and New Haven, 2000
EM—Essay on Man, Yale University Press, New Haven and London, 1944

Paul Crowther

KS—The Kantian Sublime: From Morality to Art: Clarendon Press, Oxford, 1989
CAP—Critical Aesthetics and Postmodernism, Clarendon Press, Oxford, 1993
AE—Art and Embodiment: From Aesthetics to Self-Consciousness, Clarendon Press, Oxford, 1993
LTA—The Language of Twentieth-Century Art: A Conceptual History, Yale University Press, London and New Haven, 1997
TTI—The Transhistorical Image: Philosophizing Art and its History, Cambridge University Press, Cambridge, 2002
PAP—Philosophy After Postmodernism: Civilized Values and the Scope of Knowledge, Routledge, London and New York, 2003
DACC—Defining Art, Creating the Canon: Artistic Value in an Era of Doubt, Oxford University Press, Oxford, 2007

POTVA—Phenomenology of the Visual Arts (even the frame), Stanford University Press, Stanford, 2009

TKA—The Kantian Aesthetic: From Knowledge to the Avant-Garde, Oxford University Press, Oxford, 2010

TPMA—The Phenomenology of Modern Art: Exploding Deleuze, Illuminating Style, Continuum, London and New York, 2012

POAV—Phenomenologies of Art and Vision: A Post-Analytic Turn, Bloomsbury Studies in Philosophy, London and New York, 2013

Introduction
Aesthetic Cognition, Cassirer, and Crowther

To create art—in whatever medium—is to change how the world appears. But how is this to be understood? Or perhaps the question should be—*is it to be understood*?

There is evidence that contemporary aesthetics has lost touch with those cognitive features that make the aesthetic and artistic worthy of philosophical study in their own right. The features in question here concern how art transforms our perceptions and/or imaginings, and, in so doing, changes our understanding of self, and our comprehension of other people and the world at large.

The focus of all this is the created phenomenal and historical structure of the artwork itself. This complex structure provides a unique form of sensuous or imaginatively intended knowledge, whose meaning can be experienced only through direct perceptual or imaginative engagement with the individual aesthetic or artistic object. The different art forms embody this knowledge in their own unique ways.

For convenience, we shall call all these concepts, relations, and phenomena, collectively, the doctrine of *aesthetic cognition*. They are the central concern of this book.

Aesthetic cognition—with different emphases (according to the individual thinker involved) was central to eighteenth-century thought, most notably Kant, and, thereafter for Schiller, Hegel, Schopenhauer, Bergson, Croce, Collingwood, and Dewey, amongst many others. Indeed, *it is the cornerstone of the rise and development of aesthetics as a distinctive area of philosophy*. Since the Second World War, however, its status in the English-speaking world has changed. We shall now consider why.

I Problems with Contemporary Aesthetics

A conference on 'The State of Aesthetics' was held in London in 2011. A flyer for the event stated that,

> A century ago aesthetics was of central concern for philosophers and psychologists alike. These days the study of the arts is, for both

> disciplines, a relatively minor enterprise; almost none of the world's highest profile philosophers specialise in the arts, and it is distinctly outside the curriculum's core . . . The art world is strongly influenced by philosophically inflected theories of the text, spectatorship, its own the institutions and the ideas of taste and connoisseurship. But the Anglo-American tradition in philosophy is regarded, where it is noticed at all, as too passive, distanced and traditional to be relevant to either art practice or art engagement.[1]

It must be significant that a conference responding to this thematic should be organized and attended by some of the most respected and influential figures in contemporary Anglo-American aesthetics. This shows that the marginalization referred to is a widely recognized phenomenon, and not just the ungrounded conjectures of an over-sensitive minority.

Interestingly, whilst admitting the fact of marginalization, the authors of the flyer do not consider reasons why this has occurred (although the conference itself, of course, considered some). We propose that the reason for this marginalization (especially in relation to the artworld) is because *the Anglo-American tradition has lost track of aesthetic cognition—those cognitive features and relations that enabled aesthetics to develop as a discipline in the first place*, in the eighteenth-century.

One sign of a move away from these issues by the Anglo-American tradition is the way in which the aesthetic itself has become something of a problematic notion. In the twentieth century, Clive Bell, Roger Fry, Clement Greenberg, and then Harold Osborne and Monroe Beardsley (amongst others) propounded varieties of aesthetic formalism. Their approaches involved a more limited sense of aesthetic cognition which emphasized enjoying the unity of the work *qua* aesthetic object, with any broader descriptive or narrative content taken as relevant only in relation to this.

However, none of these approaches consider the complex transformations of the relation between subject and object of experience that are basic to aesthetic cognition. Indeed, formalism is a somewhat narrow method considered in terms of both traditional and modern art, and it faces the most emphatic problems when addressing conceptual works. This is why it finds few major advocates in contemporary Anglo-American aesthetics. Even those who are sympathetic to it tend to address the statements we make about beauty and aesthetic qualities, rather than the ontology of those qualities, and what it is about them that affects us so powerfully.[2] There have also been influential theories—such as George Dickie's that are sceptical about the very notion of the aesthetic.[3]

Other developments are also implicated in the marginalization of Anglo-American aesthetics. One of its most popular topics is the analysis of the semantic and syntactic structure of different artistic media. Nelson Goodman's *Languages of Art* was a pioneering work in this area. And since that time there have been impressive analyses by Richard Wollheim, Dominic

MacIver Lopes, Peter Lamarque, and many others addressing the nature of pictorial representation and of fiction, respectively.[4]

However, this emphasis has two unfortunate side effects. The first is that the analysis of a medium's formal semantic characteristics tends to concentrate philosophical interest, on how we 'read' the medium in question. The question of understanding the mode of representation *qua* representation, rather than how we are aesthetically engaged by it as art becomes the main interest. Insofar as the question of the power of artistic meaning to change our relation to self and world is addressed, it is so under the rubric of 'expressive qualities'.[5]

This presents its own difficulties. Happiness, sadness, pity, exhilaration, joy, melancholy, and a myriad other expressive terms, all find application to art. However, what is striking is the banality of such terms when applied to art. Perhaps Brunelleschi's Dome at Florence Cathedral moves us deeply. But if it does so, it is not because it has qualities such as joy, happiness, and the like. These are almost trivial descriptive terms when applied to anything other than actual states of someone's psychological being, or to artworks that are meant only to entertain us (soap operas and the like). We might describe Beethoven's Ninth Symphony as exhilarating, but even this tells us very little. Exhilaration does not involve being blown up like a balloon with positive feelings. It is something much more complex, built upon complex networks of appraisal that link that which is being appraised to the totality of our existence. At the heart of all this is how the individual work is created and its relation to other such works. The Anglo-American tradition tends to use expressive qualities to explain why the arts matter to us. Aesthetic cognition seeks to understand what makes such properties expressive in the first place.[6]

Another feature which Anglo-American aestheticians frequently invoke in relation to the artistic aspect of the artwork, is the question of the artist's intentions. The literature on this is enormous.[7] However, the problem of complexity again arises. The relation between an artist and his or her intentions is profoundly influenced by the process of designing and making a work itself. There may be some basic intention or set of definite intentions that lead an artist to undertake the creation of a work, but the process of carrying this out brings changes almost inevitably, because it is only through the creative process itself that the work emerges.

One reason why the *making* of art has been so downplayed in contemporary Anglo-American aesthetics is because of its tendency to use Duchampian 'ready-mades' as a central example of an artwork. In such works, the question of the artist's intentions can be focussed on without any complications arising from the process of making. Any definition of art worth its salt has to accommodate such works, but when this is done through Institutional definitions such as Danto's and Dickie's (and related approaches such as Levinson's) this means that non-exhibited properties—issues concerning cultural settings rather than the phenomenal features of artworks themselves—become the focus of consideration. The way that artworks change the world and our sense of self through the individual ways they are made, becomes 'old hat'.

However, the present writers argue that—between basic semantic and syntactic structure, expressive qualities, the artist's intentions, and the experience of the audience—there is a continent that has become lost to the Anglo-American tradition. It is aesthetic cognition in the sense outlined at the beginning. Only the cashing-out of those cognitive issues and relations embodied in it will explain the expressive significance of the arts, and the relevance (or irrelevance) of the artist's intentions.

Now, whilst the authors of 'The State of Aesthetics' flyer noted earlier admit the fact of the marginalization of aesthetics, they nevertheless think that things are 'looking up'. This is because

'Recent issues of the British Journal of Aesthetics contain articles by leading figures in philosophy not usually associated with research into the arts, articles on aesthetics are once again appearing in the best generalist philosophy journals, and the trend in aesthetic writing towards making good use of theories and distinctions in philosophy of mind, philosophy of language and metaphysics continues. There is also a growing list of publications in aesthetics from such unexpected directions as neuroscience and evolutionary psychology-developments not universally welcomed within the aesthetics community.'[8]

The signs of hope, then, are aestheticians publishing in general philosophical journals and making use of ideas from philosophy of mind and metaphysics, and contributions to aesthetics from other disciplines such neuroscience and psychology. However, there are grounds for unease about such developments. The fact they are happening could be further confirmation that aesthetics is narrowing its scope and focussing on secondary problems mainly of interest to other areas of philosophy and disciplines (or, at least especially amenable to them), instead of those problems that are historically and conceptually central to aesthetics itself. And this carries the further risk, that by importing ideas and methods from these other areas, aesthetics will gradually lose the philosophical vocabulary and techniques to even address the true problems. The theory is that such developments will revitalize aesthetics; we say that they will probably exacerbate the marginalization.

There is one startling example of this. 'Aesthetic testimony' is fashionable concept amongst some Anglo-American aestheticians.[9] It is the idea that to evaluate or recognize the aesthetic worth of some object does not require that one has directly perceived it; we can make such judgments on the basis of reliable testimony from others. The way this position is presented tends to be in the manner of philosophy of mind and language. In particular, the emphasis on testimony and 'aesthetic belief' as concepts are imports from philosophy of mind. More generally still, positions in aesthetics are often assimilated nowadays under the realist/anti-realist labels employed in epistemology and metaphysics.

However, the very reason why the aesthetic and the artistic were theorized in the first place is because of their unique nature. The uniqueness here is not just logical, it encompasses how the work changes the appearance of things and transforms experience in ways that can only be recognized through direct perceptual acquaintance or imaginative engagement with the work or phenomenon in question.

This is a conceptual relation. If something is worthy by virtue of the way it is configured as sensuous or as an imaginatively intended particular, then it is logically presupposed that this worth can only be recognized through knowledge by acquaintance. It may be that we see photographs of pictures and make judgments on the basis of them; it may be indeed, that we can construct a holographic facsimile that is perceptually indiscernible from the original. However, even here, we cannot rely on these in making our aesthetic judgments unless we have perceived the original ourselves. Without this acquaintance we cannot know how faithful the reproductions are as records of the originals.

It seems to us that the main reason why the concept of aesthetic testimony has been given any mileage is because the emphasis is tacitly placed on the notion of testimony rather than on the aesthetic, and because some conceptual works—such as Duchamp's ready-mades—seem to have artistic significance that can be sufficiently warranted just by describing them second-hand. But if this is allowed, it means that the notion of the aesthetic simply means that which has to do with describing artworks. The notion of the aesthetic, in other, words, becomes so broad as to have no real significance.

If our analysis of this is correct, it means that the Anglo-American tradition's adaptation of problems in aesthetics to the methods and interests of other areas of philosophy and other disciplines, is actually one of the marginalizing factors, and not at all a solution to it.

It might be objected that these factors are not marginalizing at all—they simply reflect the way that the Anglo-American tradition has had to adapt to changes in art itself. The rise of realist and modernist tendencies in all the arts from the nineteenth-century rendered the phenomena of aesthetic experience and artistic creation much more complicated, and required new theoretical frameworks that go beyond those established in the late eighteenth-century. In particular, the Conceptualist tendencies arising from Duchamp, and which have dominated art since the 1980s make the old approaches entirely outmoded.

However, the purpose of aesthetics is not to ratify a particular phase in the history of visual art, but to understand the basis of aesthetic experiences and art on a universal basis—encompassing all the arts of the past, including those practiced outside the Anglo-American sphere. Indeed, it should be emphasized also that conceptualism and its parallels in other arts such as literature, music, and cinema, is by definition, a fringe activity that arose to criticize the nature of art as practiced and understood. The fact that—in the visual arts, at least—it became something of an orthodoxy from the late 1980s to the first decade of this century means that it has moved away from its critical stance. This means that it may now be more amenable to consideration in terms of concepts central to aesthetics as such, rather than through concepts especially adapted to it. (We will return to the question of Conceptual art in more detail further on in our book.)

Now, it is important to reiterate that in the foregoing analysis we have not been intending a global critique of Anglo-American aesthetics, as such. It has had some great achievements. We have been seeking, rather, to register grounds for unease as to its contemporary status. Specifically, we fear that most of its best achievements have focussed on issues that are secondary to the big questions concerning aesthetic cognition, and have actually deflected attention away from these. This why Anglo-American aesthetics has become marginalized from the role played by the arts in our broader culture. Further, the process of adopting ideas and methods from other areas of philosophy and from other disciplines, threatens to water down the aesthetic concepts which would help it become more culturally relevant.

There is one feature still to be considered which does constitute a genuine criticism of Anglo-American aesthetics as a method, per se. We will defer consideration of it, however, until after we have considered aesthetics in the tradition of Continental philosophy. Some important thinkers from this tradition will be briefly addressed, in turn, noting difficulties that are particular to them, and then some more general concerns will be considered.

First, Heidegger. His essay 'The Origin of the Work of Art' has been profoundly influential.[10] In it, he argues that we should regard the artwork as a bringing forth or disclosure of Truth. The Truth in question here, is that of the Being of being—the 'isness' of that which is represented in the work. Heidegger considers examples including a painting by Van Gogh of (allegedly) peasant shoes, a Greek temple, and poetry by Holderlin.

The kind of disclosure Heidegger describes is of extreme importance. In many respects art does indeed emphasize the being of that which it represents. However, Heidegger's account is extremely one-sided and more or less *appropriates* artistic creation to the demands of his own philosophy. This means that other key features of the artwork are marginalized, or even overlooked entirely. We could mention, for example, that no attention is paid to the importance of the individual style through which the artist is able to disclose the important features of the subject matter. Neither does Heidegger consider the broader ways in which our perception of this style allows to empathize with the artist's way of presenting things. One wonders also, what distinguishes the presentation of truth in painting from its presentation in poetry or architecture. These individual media must have their own distinctiveness, but Heidegger does not appear to negotiate this in any depth. (Neither, for that matter, does he consider what distinguishes the disclosure of truth in art from its disclosure by philosophy.)

In the case of Merleau-Ponty's influential philosophy of art some attention is paid to the unique character of the different media.[11] However, like Heidegger, he assimilates art very closely to the terms of his own philosophy. Because this philosophy emphasizes the centrality of embodiment in perception, he tends to privilege painting. He argues that it discloses the role of the body in perception at a primordial level—where the recognizable perceptual world begins to take on form. Painting discloses the normally unnoticed or 'invisible' features and relations that allow things to become visible.

Again, there is no doubt that the cognitive factor which Merleau-Ponty emphasizes is an important part of why we find painting meaningful and of value. However, his aesthetics is too orientated towards painting, and even in relation to it, he has some of the same drawbacks that Heidegger has. For example, the role of the artist's individuality is not well explained, and there is little attempt to explain the nature of our aesthetic experience of art. It is also the case that he does not pay enough attention to the way in which paintings transform how things appear. He emphasizes how they disclose 'invisible' features that are present in vision even when not explicitly noticed. But he does not consider how painting changes the appearance of the visible and the 'invisible' in relation to structures of the medium itself.

In recent years, the work of Gilles Deleuze has become very influential in Continental aesthetics. The present writers find some of his work obscure to the point of unintelligibility. However, his book on Francis Bacon presents a masterly analysis of painting as such as well as Bacon in particular.[12] Deleuze emphasizes the importance of 'Figure', which is what we attend to when considering how the marks in a painting emerge from the ground to constitute forms at a higher level.

Unfortunately, as Crowther has argued in great detail, Deleuze makes this into the basis of an essentialist hierarchy.[13] In Deleuze's terms, true artists of the Figure are those who work in a coloristic idiom that juxtaposes areas of color to create impacts that we experience as 'sensations'. However, the way Deleuze articulates Figure means that it should be available, in principle, from any painting that is good enough to make us look at how it works, pictorially speaking. We cannot privilege one artist over another in relation to this on the basis of the kind of style he or she has. It must be done on the basis of the artist individually considered.

There are a number of other thinkers who have been very influential in Continental aesthetics. Lyotard, Nancy, and Ranciere, for example, have popularized a kind of art/philosophy journalism, which emphasizes discussion of the cultural status of images rather than the nuances of their cognitive richness.[14]

All the thinkers share some broader problems. With the exception of Ranciere, none of them have a well-developed notion of the aesthetic, nor, for that matter, of art *qua* artistic. There is also a more general and profoundly important worry that applies also to most Anglo-American aesthetics. Some thinkers in both traditions have linked the aesthetic and artistic to cognition, but they have understood this very much in terms of perception of the object, rather than as a factor that connects the subject to it—in a way that changes its understanding of both. This enhanced interaction of subject and object is at the heart of what we have called *aesthetic cognition*. However, it involves a feature that has not—with one major exception—been given its due by either tradition, namely that of its historical dimension.

The exceptional philosopher just referred to is Hans-Georg Gadamer. He affirms the importance of 'play' as a clue to the ontology of the artwork. The symbolic character of the work illuminates the totality of our relation to the world, and allows us to experience in an enriched way—an enrichment that has historical meaning. However, for the present writers, Gadamer does not adequately connect this to the artist's activity in engaging with and transforming the tradition in which he or she works. He admits that 'When someone makes an imitation, he has to leave out and to heighten. Because he is pointing to something, he has to exaggerate whether he likes it or not'.[15] As we will show later, this is the very heart of the matter. It is the artist's style in creating the image, considered in relation to broader factors concerning the role of imagination in experience, and in relation to the history of the medium in question, that is the basis of aesthetic cognition. Gadamer takes us to the threshold of the problem, but then explores it with a concept of historical understanding that is mainly derived from his own broader philosophical hermeneutics. Our point is that the sense of history involved has to be made more specific to the arts and also made much more medium-specific.

The matter can be put in outline like this. When people make art, it changes things. It does so, because the work discloses the world in a way that attracts attention. It can do this because it is not mundane or formulaic in its approach. It stands out from other works in the medium that presently exist, or which have been created before. In order to do this, it does not have to be innovatory in any deep sense, but when it is, it has profound ramifications for how other artists create things, and that mean for how the audience perceive the world. The point is that cognition in and through art is mediated by a comparative historical horizon. When we experience something that transforms how things appear or are experienced, we recognize the transformation through its difference from the way things are usually done in that respect, within the traditions of that medium.[16]

This can be considered under a different aspect. No art is created in an absolutely private context. We can write poems alone in our room and show them to no one, but learning how to write them and the desire to do so, are features absorbed through learning how others have done such writing. Once an artwork is created in a publically accessible context it changes the nature of how the medium of which it is an example. When we recognize positive difference in terms of what has gone before, it becomes a vehicle for changing what is to come in that medium. Other artists can learn from it, put it to use, change it, or whatever.

These points have a key implication. The arts can progress; they can advance. There can be a canon. The comparative historical dimension of aesthetic cognition in the arts is not static, it engages with the present by positioning itself in relation to the past and by hoping to create an influential future. The importance of all this was recognized in very different ways by Hume, Kant, and Hegel, but has largely dropped from view as a vital factor in contemporary aesthetics.

We have outlined, then, some reasons to worry about the present state of aesthetics in both the Anglo-American and Continental traditions of aesthetics. The former no longer connects with aesthetic cognition, and the latter does only in very restricted one-sided ways. Neither tradition does much justice to aesthetic cognition's historical dimension.

It might be objected that the life of the arts in contemporary culture is very diverse, and that—in the Continental tradition, thinkers such as Foucault, Lacan, Derrida, and Deleuze have emphasized the importance of difference, heterogeneity, and the decenteredness of the self. In such a context, a foundational approach in aesthetics is no longer viable. We do not accept this. Indeed, we propose a *refoundational* approach. This is one that links art and the aesthetic to constants in experience that are functional rather than substantial, i.e., ones whose structural principles are realized differently in different historical contexts.

To accomplish this task we draw on Ernst Cassirer and Paul Crowther. Cassirer points the way.

> The creative process must always satisfy two different conditions: on the one side, it must tie itself to something permanent and existing, and on the other, it must always be open to a new employment and a new approach that changes this existence. For only in this way does one succeed in doing justice to the demands of both the object and the subject.[17]

Cassirer gives a rich theory of the epistemological and historical basis of symbolism in the relation between subject and object of experience, and in aesthetics as a historically developing discipline; and Crowther offers the means of thinking this through in relation to contemporary aesthetics and different art media. The upshot is a dynamic theory which restores aesthetic cognition to its warranted centrality. In particular, the Cassirer-Crowther link allows us to understand art as a direct continuation of the most primal level of signification. We shall now explain a little more about our choice of these two thinkers and our reasons for linking them.

II Cassirer and Crowther

Ernst Cassirer's philosophy of symbolic forms is a major contribution to twentieth-century thought and interest in it has been undergoing considerable revival recently. There is now an acute and burgeoning body of exegetical work on him, led by Donald P. Verene's pioneering work, and important monographs by John Krois, Thora Ilin Bayer, and S.G. Lofts.[18]

Cassirer's way of approaching symbolic structures considers both the formal features of signification, and the ways in which these are concretely embodied in our historical existence. He shows how different areas of human experience are symbolically structured in logically distinctive but interacting and developing ways. The basis of Cassirer's philosophical position is found in the following remarks.

> Only where we succeed, as it were, in compressing a total phenomenon into one of its factors, in concentrating it symbolically, in "having" it in a state of "pregnance" in the particular factor—only then do we raise it out of the stream of temporal change; only then does its existence which had hitherto seemed confined to a single moment in time, gain a kind of permanence: for only then does it become possible to find again in simple, as it were, punctual "here" and "now" of present experience a "not-here" and a "not-now". Everything that we call the identity of concepts and significations, or the constancy of things and attributes, is rooted in this fundamental finding-again.[19]

In the present book we shall call this *the basic symbolic function*. Its formal structure is the expression of a kind of urge to create meaning that is grounded in our existence as embodied beings. As Cassirer says,

> Since every particular content of consciousness is situated in a network of diverse relations, by virtue of which its simple existence and self-representation contain *reference* to other and still other contents, there can and must be certain formations of consciousness in which the pure form of reference is, as it were, sensuously embodied.[20]

This concern with this intimate link between perception and symbol anticipates the concerns with theories of meaning and of 'signification' that have come to dominate recent tendencies in Anglo-American and Continental philosophy, respectively. Cassirer offers a way of negotiating between these. At the same time, as a representative of the philosophical tradition of Kant and Hegel, he offers, as well, a way of recontextualizing the debate about meaning and perception.

Cassirer realized that understanding art as a symbolic form was of potentially great philosophical importance, and he hoped to write a book on the topic. This was never done. He wrote a few relevant papers, and considered Enlightenment aesthetics in some detail, but was only able to write one substantial piece. This is the chapter on art in *An Essay on Man* (1943), where he gathers up the most important ideas from his earlier papers and develops them further. By doing this, Cassirer creates an opportunity for not only understanding the symbolic basis of art, but also seeing it as a unique expression of our general cognitive relation to the world. In the following remarks he conveys the basis of art as a symbolic form and as an expression of aesthetic cognition.

> No artist can represent nature without, in and through this representation, give expression to his own I; and no artistic expression of the I is possible without setting before us something objective, feeling and gestalt must merge into one another and must be completely absorbed in one another if a great work of art is to be created. However, it follows from this that—and why—the work of art can never be a mere production of the subjective or of the objective, of the mental or the objective world, but that a genuine discovery of both takes place here: a discovery that, in its general character, falls short of no theoretical knowledge.[21]

The best introduction to Cassirer's understanding of art is a very fine chapter in S.G. Lofts's monograph.[22] There have also been some interesting attempts to develop individual aspects of Cassirer's aesthetics.[23] Heinz Paetzold used Cassirer's ideas for constructing an outline philosophy of culture based on symbolic form.[24] Susanne Langer's celebrated *Feeling and Form* proclaims a debt to Cassirer, and we shall make some use of her notion of semblance later on.[25] In practice, however, the link between her ideas and the specifics of Cassirer's position is much more tenuous. In particular, she does not address the profound connection between art, imagination, and the unity of self-consciousness that is posited by Cassirer, and which will be a focus of the present work.

The only dedicated study that negotiates Cassirer's aesthetics in depth, and on its own terms is a monograph by Marion Lauschke.[26] It is an excellent analysis of the origins of Cassirer's thinking about art and the aesthetic in German idealism and beyond. However, its formulation of art as a symbolic form lacks emphasis on how the basic symbolic function is uniquely embodied by the different art media. In effect, Rauschke offers a theory of art as a symbolic form that fills in some further details of Cassirer's general position, but does not address how the theory is to be made more ontologically and historically concrete as a working concept. These features will be central to the account offered in this work.

There is one other feature of Cassirer's aesthetics that Langer and Rauschke do not give due consideration to. It is the question of *progressive articulation*. This phrase was devised by Paul Crowther to mean

> 'a process of cumulative and sometimes erratic advancement where each form of knowledge is able to build on its previous stages so as to become more internally complex and detailed and more comprehensive in its explanatory and/or methodological scope.[27]

This perfectly describes Cassirer's own understanding of the historical development of symbolic form. In terms of the arts, they develop historically in a process that amounts to more than change as such. Cassirer realized also that this is also true of thinking about the arts, i.e., aesthetics as a discipline, itself. Any aesthetic theory operates historically whether it is aware of it or not, building upon and responding to previous approaches, even if only through inheriting established attitudes.

Contemporary aesthetics, as it stands, does not appear to us to have either the conceptual means for negotiating the reciprocity between the unique symbolic structures of art media and their individual creative articulations, or the role of historical positioning in practice of the arts, and in aesthetic theory. Cassirer's philosophy and his conception of aesthetics offer a starting point for negotiating both.

However, in order to deliver on this promise, his approach has to be both considered and developed further. We do not presume to offer some systematic extension of the philosophy of symbolic forms. Our intention rather is to follow some of his main insights and to show the enduring potential of his work. However, we also feel that the importance of Cassirer's work for understanding art and culture can be developed most effectively by linking it to Paul Crowther's ideas.

Crowther is, in many respects, an enigma. He is the most prolific writer on philosophical aesthetics in the English language, but his output is not matched by corresponding influence. In this his work seems to fall between two stools. He explicitly attempts to integrate the Anglo-American with the Continental traditions of existentialism and phenomenology. However, neither tradition seems to have any great desire for such integration, so he is neglected by both.

This is unfortunate because Crowther is the one philosopher who has integrated the cognitive and perceptual/imaginative dimension of the arts, with the dimension of comparative history. It is notable that, in developing this aesthetic cognitivism, he has already made use of Cassirer in some detail (in *The Language of Twentieth-Century Art: A Conceptual History*, and *Philosophy After Postmodernism: Civilized Values and the Scope of Knowledge*). It goes without saying that Crowther's other nine monographs (all on aesthetics) and his important papers on imagination also furnish us with important material for understanding the contemporary potential of aesthetic cognition.

This link with Cassirer is of great use in bringing out the general importance of Crowther's work in aesthetics and beyond. One reason why his work is not studied in more detail by the Continental tradition is that he is seen as just an aesthetician. The other major Continental thinkers, in contrast, are seen as more complete through being able to link aesthetics to more comprehensive philosophical positions.

Now Crowther has set forth the glimmerings of such a general position, but, when linked to Cassirer, the two together offer a wonderful synergy of philosophical insights, which we begin to explore in the present book.[28] A central feature is the importance of imagination. John Lechte has argued that 'even if the forms of the image change, there is still something essential about the image.'[29] As Lechte presents it, this essential feature is not at all something rigid and timeless. It allows the image different ways of being realized and made meaningful under different historical circumstances. By linking Cassirer and Crowther, the intimate link between cognition and imagination that is the essential basis of the image can be understood. We can then understand how this link is embodied in the creation of art, and, indeed, the development of symbolic forms.

In the longer term, our study might also be useful as the groundwork of a broader approach that integrates Cassirer's treatment of the sign and symbolic forms with Crowther's theory of cognition and the civilizing process. This has the potential to result in a *philosophy of culture* with the most far-reaching epistemological and metaphysical implications.

The present work, then, takes (in effect) an important first step towards such a philosophy by showing how Cassirer's and Crowther's aesthetics—considered in conjunction—can revitalize our thinking about the aesthetic and artistic. This will be achieved through clarifying the role of the aesthetic and artistic in the bigger picture of our general cognitive involvement with the world, and the historical setting of this involvement.

III The Stages of Argument in This Book

Chapter 1 presents the key concepts that are the basis of Cassirer's philosophy. We elaborate the basic symbolic function, and argue that Cassirer's neo-Kantian 'critical idealism' has a flexibility that allows it to be understood as a mode of corporeal phenomenology with some kinship to Merelau-Ponty's position. However, Cassirer understands the importance of reflective knowledge in a way that Merleau-Ponty does not.

We then, discuss Cassirer's notion of symbolic form, and the importance of progressive articulation. We also dwell on the importance that Cassirer assigns to myth as a symbolic form, and its important links to imagination and the unity of self-consciousness. However, we note also that whilst Cassirer emphasizes the importance of imagination, he does not have a theory of his own, but rather tends to follow Kant's. We suggest that Crowther's theory of imagination and the unity of self-consciousness is much more useful for developing the strength of Cassirer's position.

In Chapter 2, we analyse Crowther's theory of imagination and its role in our correlated knowledge of objects and of the self; we also analyse his highly original arguments concerning the aesthetically grounded narrative unity of the self. Crowther argues that imagination is the cognitive capacity that enables trans-ostensive reference. If we are to learn language and concepts we must know that these can apply to things and situations other than ones given immediately in perception. Through gesture and habit in our ontogenetic development, we gradually correlate sign and imagination to create concepts.

Our account addresses also, the way imagination does this, through consideration of aspects of Crowther's phenomenology of the image, which emphasizes its schematic, unstable and creatively plastic character—features which mean, collectively, that the image is a stylized, quasi-sensory representation of possibility. It is also something that is employed to great effect, in memory. All this means that imagining something is a way of interpreting it on the basis of who we presently are and the interests which define us. In this way, imagination has a further vital role in the unity of the self. However, the creative element that is intrinsic to it means that this unity is more than a linear continuity, it involves creative narrative, and aesthetic considerations. This means that, intrinsic to the self—if material circumstances allow, there is a motive for self-development, and that the emergence of different symbolic forms and their progressive articulation, is, in part, an outcome of this.

Our interpretation of Crowther suggests that what we called aesthetic cognition at the beginning of this book has its origins in more basic levels of knowledge and self-awareness that center on the role of imagination. However, humanity's attempts to express this aesthetic dimension and give it advanced and publically accessible expression are always connected to the broader cultural context—especially religion and ritual. The emergence of Ideal beauty and the cultural importance assigned to the arts from the Renaissance onwards are signs of the aesthetic gradually being recognized as a unique mode of knowledge in its own right, but it is only in Enlightenment thought that what we have called aesthetic cognition begins to be articulated explicitly.

In Chapter 3, accordingly, we turn to Cassirer's detailed investigation of the rise of aesthetics as a discipline of inquiry during the Enlightenment, and its first explicit engagement with the factors involved in aesthetic cognition. He identifies a classicist-objectifying trend, and a more empiricist reception-oriented reaction to it. In Shaftesbury and Burke, he further identifies a pair of approaches which break down the barriers between objectivist and empiricist approaches by linking aesthetic experiences to a more active and formative conception of the self.

We then go on to consider how Cassirer traces the emergence of imagination as key feature in aesthetics, and the new conception of beauty—emphasizing its sensuous grounding—that emerges with Baumgarten. Cassirer's interpretation of these developments is then critically reviewed, and we note, in particular, some difficulties in his interpretation of Burke.

We then present Cassirer's important and original reading of Kant's aesthetics and the particular importance that he assigns to formal structure, and the universal validity and disinterestedness of the pure aesthetic judgment. A critical review of Cassirer's interpretation of Kant is then offered. We argue that difficulties arise through his neglect of Kant's concept of the 'aesthetic idea' and his interpretation of the pure aesthetic judgment's claim to universal validity. Ideas from Crowther are used to clarify his problems here.

Chapter 4 addresses mainly Cassirer's important chapter on art from *An Essay on Man*. We consider his arguments that art is a formative activity that addresses both the object-orientated and subjective aspects of human existence. Some difficulties on Cassirer's part are noted, and dealt with provisionally. Consideration is then given to Cassirer's account of the development of the concept of imagination in aesthetics. In effect, he shows us how this concept becomes progressively articulated as aesthetics as a discipline develops. We also address his arguments concerning the limitations of other aesthetic theories contemporary with his own work (or from the decades immediately preceding it).

Towards the end of the chapter, we show the limitations of Cassirer's theory especially vis-à-vis the relation between subject and object of experience in aesthetic cognition. However, these limits can be overcome by developing his position through using concepts from Crowther. In particular, Crowther's normative definition on art—with its emphasis on originality, allows the comparative historical dimension to be assigned due role in aesthetic cognition.

The main significance of Chapters 3 and 4, then, is in showing how Cassirer, in effect, extends the philosophy of symbolic forms by explaining the birth of aesthetics as a discipline in relation to concepts and relations involved in aesthetic cognition. There is the gradual realization that art and the aesthetic are the basis of a unique symbolic form that engages with some of those deep levels of experience that were described in Chapter 2. Cassirer's account, in other words, is one that shows the correlation between aesthetics as a progressively articulating discipline and aesthetic cognition in a broad sense. It is this continuingly relevant dynamic that the Anglo-American tradition has lost sight of and the Continental tradition engages with only through narrow aspects.

However, for Cassirer's project to be made fully relevant it has to be upgraded. Throughout Chapters 3 and 4, we make use of ideas from Crowther's general aesthetics to do this. However, there is one particular issue that has to be developed further. Cassirer constantly states that within art as a general symbolic form, the individual art media each have their own unique features. But he does not explain the basis of this in any detail, or what broader effects might follow from this. Crowther's work has its greatest strengths in exactly this direction.

In Chapter 5, accordingly, we look at his philosophy of the visual arts. First, Crowther's account of the cognitive basis of architecture and its important relation to the body are presented in detail. We then show how the development of an outstanding personal style can progressively articulate the symbolic scope of a medium through both technical and compositional achievement. As a case history, we look at the example of Brunelleschi, and in particular his work on the Dome of Florence Cathedral. We then adopt, in turn, a similar strategy in relation to pictorial art, photography, digital art, and finally Conceptual art. Our case histories are, respectively,

Raphael and Picasso for pictorial art, Alfred Stieglitz for photography, William Latham for digital art, and Marcel Duchamp for Conceptual art.

The Conceptual art discussion is of particular importance. Crowther's general strategy shows how aesthetic cognition is not something that has been rendered obsolete by Conceptualism. Many Conceptual works, indeed, require consideration in such terms for their full potential to be appreciated. Our use of Duchamp's artistic development hopefully brings this out in full.

Chapter 6 considers some of the major arts of temporal realization, namely, literature as such, poetry, storytelling, theatre, and music. Of particular importance is our development of Crowther's explanation of the origins of literature in rhetoric, and the use of devices such as metonym and metaphor. We then follow the same strategy as in Chapter 5, developing ideas from Crowther to identify the features that are unique to the respective media, and then using examples to show how an individual's stylistic development can lead to the relevant medium itself being progressively articulated. As case histories for the latter, we use Mallarme, Cervantes, Aeschylus and Brecht, and Steve Reich.

Finally we offer a conclusion that summarizes the main argument of our book as a whole. This argument concerns the way aesthetic cognition originates in the basic symbolic function and the complex role of imagination, and is then embodied through artistic creation and aesthetics (as both experience and understanding) and is, at the same time, progressively articulated. We end with some important considerations of the universal significance of aesthetic cognition, and its importance for education and the development of critical autonomy, and philosophy of culture.

Notes

1 The flyer was circulated electronically through Stephen Clarke's online philosophy list <PHILOS-L@liverpool.ac.uk> in early 2011.
2 See, for example, Nick Zangwill, *The Metaphysics of Beauty* (Ithaca: Cornell University Press, 2001).
3 See, for example, his intense critique of Kant in *The Century of Taste: The Philosophical Odyssey of Taste in the Eighteenth Century* (Oxford: Oxford University Press, 1996).
4 See, for example, Richard Wollheim, *Painting as an Art* (London and New York: Thames and Hudson, 1987); Dominic M. Lopes, *Understanding Pictures* (Oxford: Oxford University Press, 1996); and Peter Lamarque and Stein H. Olsen, *Truth, Fiction, and Literature: A Philosophical Perspective* (Oxford: Oxford University Press, 1996).
5 In what follows, we are using the critique of Wollheim's emphasis on expressive qualities and the artist's intentions that is found in Chapter 1 of *POAV*. The points Crowther makes there are also relevant to all the major expressive qualities/intentions theorists.
6 There are, of course, some who take a much more useful approach such as Matthew Kieran in his *Revealing Art: Why Art Matters* (London and New York: Routledge, 2004). Kieran picks up on some features of what we call aesthetic cognition,

but is only able to apply these to particular artworks in an ad hoc way. Admirably, however, he does have some important insights concerning the importance of the artist's engagement with the medium as a feature in the aesthetic worth of the artwork. Of all the contemporary thinkers in Anglo-American aesthetics, Kieran is the one with most affinity to Crowther. It would be interesting to do a comparison and contrast of their approaches in relation to the visual arts in particular.

7 A very comprehensive treatment of Anglo-American approaches can be found in Paisley Livingston's *Art and Intention: A Philosophical Study* (Oxford: Clarendon Press, 2007).

8 Clarke list, op. cit.

9 See, for example, Aaron Meskin's "Aesthetic Testimony: What Can We Learn from Others About Beauty and Art?", *Philosophy and Phenomenological Research* 69, no. 1 (2004): 65–91.

10 The essay is conveniently available in Martin Heidegger, *Basic Writings*, 1st Harper Perennial Modern Thought Edition, ed. David Farrell Krell (New York: Harper Collins, 2008), 143–212.

11 See especially, the essay 'Eye and Mind', included in *The Primacy of Perception*, ed. James E. Edie, trans. Carleton Dallery (Evanston, IL: Northwestern University Press, 1964), 159–190.

12 Gilles Deleuze, *Francis Bacon: The Logic of Sensation* (Minneapolis: University of Minnesota Press, 2005).

13 See Crowther, *TPMA*, pp. 14–56. The kind of worries that we raise in relation to contemporary aesthetics are also raised by him in in the context of visual art through his detailed critiques of Wollheim, Heidegger, Merleau-Ponty, and Lacan in the relevant Chapters of *POAV*.

14 See, for example, Jean-Francois Lyotard, *The Inhuman*, trans. Geoff Bennington and Rachel Bowlby (Cambridge: Polity Press, 1991); Jacques Ranciere, *The Future of the Image*, trans. Gregory Elliott (London and New York: Verso, 2007); Jean-Luc Nancy, *The Ground of the Image*, trans. Jeff Fort (New York: Fordham University Press, 2005).

15 Hans-Georg Gadamer, *Truth and Method*, trans. William Glen-Doepel (London: Sheed and Ward, 1979), 103.

16 This theory is the basis of Crowther's normative aesthetics which we use throughout this book. Its best statement is in *DACC*.

17 *LCS*, p. 116.

18 Donald Phillip Verene's most recent relevant monograph is *The Origins of the Philosophy of Symbolic Forms: Kant, Hegel, and Cassirer* (Evanston, IL: Northwestern University Press, 2011); John M. Krois, *Cassirer: Symbolic Forms and History* (London and New Haven: Yale University Press, 1987); Thora I. Bayer, *Cassirer's Metaphysics of Symbolic Forms* (London and New Haven: Yale University Press, 2001); S. G. Lofts, *Ernst Cassirer: A "Repetition" of Modernity* (Albany, NY: State University of New York Press, 2000). A very useful collection of essays (especially the Editor's own introduction) can be found in Jeffrey A. Barash, ed., *The Symbolic Construction of Reality: The Legacy of Ernst Cassirer* (Chicago and London: University of Chicago Press, 2008). A good general Introduction to Cassirer's thought is Robert Skidelsky's *Ernst Cassirer: The Last Philosopher of Culture* (Princeton: Princeton University Press, 2008).

19 *PSF*, Vol. 3, p. 114.

20 *PSF*, Vol. 1, p. 106 (See also *EM*, pp. 225–226).

21 *LCS*, p. 31.

22 Lofts, *Ernst Cassirer. . .*, pp. 183–199.

23 Other recent works that make use of aspects of Cassirer's aesthetics include Carmen Metta, *Forma e figura. Sul problema della rappresentazione in Ernst Cassirer e Paul Klee* (Macerata: Quodlibet, 2009); Peer F. Bundegaard, "The

Grammar of Aesthetic Intuition: On Ernst Cassirer's Philosophy of Symbolic form in the Visual Arts", *Synthese* 179, no. 1 (2011): 3–17; and Thora I. Bayer, "Art as Symbolic Form: Cassirer on the Educational Value of Art", *The Journal of Aesthetic Education* 40, no. 4 (Winter 2006): 51–64. Metta's work is of short book length, but links Cassirer to Klee on the basis of ideas from the latter that are, in themselves, highly problematic.

24 See, for example, Heinz Paetzold, *The Symbolic Language of Culture, Fine Arts and Architecture: Consequences of Cassirer and Goodman* (Trondheim: FF Edition, 1997). In this book, Paetzold links Cassirer and Goodman. It is true that Goodman's major works in the field of aesthetics offer cogent and provocative analyses of the symbolic structure of different art media, and (without elaborating much on the connection) Goodman himself has also noted some affinities between his approach and Cassirer's Philosophy of Symbolic Forms. However, these affinities are too general to be of much use. And there are more specific disaffinities of a serious kind. Goodman, for example, is very much viewer-orientated in his approach to art whereas Cassirer—as we have already shown—stresses its formative significance as an act of producing symbolic meaning. Related to this is the fact that Goodman finds little space for art's historical dimension, i.e., its possibility for progressive articulation. But for the purposes of the present book this dimension of meaning is of the greatest significance.

25 Susanne K. Langer, *Feeling and Form: A Theory of Art* (London: Routledge and Kegan-Paul, 1953).

26 Marion Laushke, *Ästhetik im Zeichen des Menschen: Die ästhetische Vorgeschichte der Symbolphilosophie Ernst Cassirers und die symbolische Form der Kunst* (Hamburg: Felix Meiner, 2007).

27 Paul Crowther, *PAP*, p. 44.

28 Crowther's general philosophical position is set out in most detail in *PAP*.

29 John Lechte, *Genealogy and Ontology of the Western Image and Its Digital Future* (London and New York: Routledge, 2012), 3.

1 The Philosophy of Symbolic Forms

Introduction

The volumes, which comprise the Philosophy of Symbolic Forms, were written between 1923 and 1929 (although the incomplete fourth volume was published only posthumously in 1996). In these works Cassirer develops a form of 'Critical Idealism'. Of his later works *An Essay on Man* (1944) is of marked interest because it not only summarizes cogently his general philosophical endeavour, but also extends it in important ways. Indeed, in relation to the aims of this book, *An Essay on Man* is of even more importance because it includes his most extensive piece of writing on art as a symbolic form.

The present chapter first considers Cassirer's epistemology and tries to put it in a workable form.

I The Basic Symbolic Function

Cassirer conceives philosophy's general task as follows:

> It must follow the special sciences and survey them as a whole. It must ask whether the intellectual symbols by means of which the specialized disciplines reflect on and describe reality exist merely side by side or whether they are not diverse manifestations of the same basic human function. And if the latter hypothesis should be confirmed, a philosophical critique must formulate the universal conditions of this function and define the principle underlying it.[1]

The task of the philosophy of symbolic forms is lucidly declared in this passage. Cassirer's goal is to comprehend the specialized structures of different forms of knowledge, relate them to one another as a whole, and through this to identify the basic formative principle which is exemplified in them all.

Cassirer follows Hegel in arguing that this unification must address the way forms of knowledge develop through their own internal logical and historical progression, and through their relations to one another. Cassirer, indeed, sees the immediate antecedents of his own project lying in a crisis of modern philosophy, which is eventually resolved by Kant. The problem is that both the rationalist and the empiricist epistemologies commit the same error. However much they differ in their explanations of how our mental representations relate to their object, they agree—far more than they realize—in their formulations of, and methodological approaches to, the problem. Both theories posit mediating acts which transform the mere datum of consciousness into an 'objective intuition', a content of reality that embodies the "outside world". According to Cassirer,

> Empiricism reduces this metamorphosis to associations and reproductions; rationalism reduces it to logical operations, judgments and inferences. But what both overlook in equal degree is the circumstance that all the psychological or logical processes, here invoked, come in a manner of speaking, too late; they all refer to a combination of elements which are viewed as in some way existing, as established before the combination. But the question involved here does not begin with the possibility and basis of combination; it is concerned with the possibility of positing the ability to combine.[2]

For both rationalism and empiricism, the concept is abstracted from data, which are presented to consciousness in basic acts of attention. However, Kant's Critical Philosophy overcomes the confusions of rationalism/empiricism by shifting the problem of the concept from ontology—as something abstracted from the given—towards epistemology.

> Instead of starting from the object as the known and given, we must begin with the law of cognition, which alone is truly accessible and certain in a primary sense; instead of defining the universal qualities of *being*, like ontological metaphysics, we must by an analysis of reason, ascertain the fundamental form of judgment and define it in all its numerous ramifications; only if this is done can objectivity be conceivable.[3]

This is a very broad notion of objectivity, that—in the *Critique of Practical Reason*, and The *Critique of Judgment*, respectively, Kant extends to include ethical freedom art and natural organic forms. Its broader implication is that knowledge refers

> no longer solely to the function of logical judgment but extends with equal justification and right to every trend and every principle by which the human spirit gives form to reality.[4]

Cassirer's task is, accordingly, to develop this basic strategy even further, and to explain the logical structure and historical development of different symbolic forms and the formative principle that underlies them. In practice, this means starting from the way in which language acquisition refines the basic formative principle and enables reference to items and relations to be achieved abstractly. Signification is achieved without having to be stimulated by the presence of that to which it is referring.

This *abstract function* (as we shall label it) can be achieved in different ways. There are different *symbolic forms*. Cassirer proposes language, myth and scientific knowledge, and art as basic symbolic forms. Later, in *An Essay on Man* he also adds history to this list. The thinking behind this is not made entirely clear by Cassirer. It is curious to regard language as one symbolic form amongst others because it is clear that whilst mathematical relations (which are essential to scientific knowledge) and modes of visual representation are not written or spoken language, they have, nevertheless, an at least quasi-linguistic character.[5]

Our overview of Cassirer's philosophy of symbolic forms is now in place. The philosophical foundations of this must now be considered in more detail— with a view to making the Kantian aspects more flexible

II Cassirer as Phenomenologist

We begin with the question of a priori features of experience. In Kant, a number of factors have 'a priori' status in that they are preconditions of any possible experience. These are space and time, and twelve categories (or 'pure' concepts of the understanding) including the causal relation and substance. Cassirer himself singles out space, time, causality, thing/attribute, number, and magnitude as having fundamental experiential significance. The question arises as to whether these have claim to a priori status, and if so, what relation they have to Kant's fundamental conceptual framework.

In *Substance and Function* Cassirer defines the a priori as follows:

> Only those ultimate logical invariants can be called a priori, which lie at the basis of any determination of a connection according to natural law. A cognition is called a priori not in any sense as if it were *prior* to experience, but because and in so far as it is continued as a necessary premise in every valid judgment concerning facts.[6]

It is fairly clear that throughout the Philosophy of Symbolic Forms Cassirer regards space, time, causality, thing, number and magnitude in these terms. Interestingly, however his use of the term a priori is almost exclusively applied to Kant's philosophy in the course of his exposition of the Philosophy of Symbolic Forms. The one major exception to this is in the final volume where Cassirer tells us that:

> The concepts and propositions that the unity of our theoretical knowledge of nature, its divisions and systematic structure, are based upon,

> are never taken simply from experience. They are logically prior to it in the sense that they represent the line or direction for thought. They set the goal for it, and they help to prepare the way that leads to this goal. The way is not prescribed from the beginning, and no markings exist; clearing the way is the true task of theoretical thought and it is in this that the achievement of the a priori consists.[7]

In this passage Cassirer is affirming his difference from Kant by underlining the dynamic nature of the conception of categorical structure in the Philosophy of Symbolic Forms. His categories form a field of a priori logical invariants, but they are articulated in different ways as human experience progresses. They are functional constants.

This would suggest that Cassirer's reticence concerning the use of 'a priori' as term is because he recognizes important differences between his and Kant's notion of a categorial field. Both agree that in logical terms without such a field no experience of an objective world would be possible. However, they both disagree on the exact contents of this field, and on how the principles, which it contains, actually work in epistemological terms. In relation to the former, for example, Kant's system of categories has always been criticized for the way it is derived somewhat artificially from Aristotle's philosophy. Cassirer's categories in contrast are much more economic and it might be possible to see the Kantian framework as derivable from them. However, the more interesting question is the contrasting epistemological structure and function, which Kant and Cassirer assign to their categorial fields.

For Kant experience is transcendentally ideal in a radical sense. Space and time, causality and other categories are *exclusively* functions of the judging subject. They are principles for organizing experience *per se* and we have no justification for interpreting them as having application beyond human experience. The epistemological status of Cassirer's categorial field, in contrast, is *much* more complex. Matters come to a head here in relation to space and time. In his book on Kant, Cassirer tells that:

> There is a definite form of objectivity for us, which we call the *spatial order* of things; we must understand and determine it, not because we examine and analyse existence of an absolute space but because we examine and analyse the laws of geometrical construction, those laws by which points, lines, planes, and bodies arise for us by way of continuous construction.[8]

This conception of space is the one which Cassirer himself affirms throughout the Philosophy of Symbolic Forms. It is not, however, the same as Kant's. For whilst in Kant the unity of space is describable in the terms which Cassirer uses, space as such is for Kant more than a mere principle of unity. It is a *form of intuition*, which is fundamentally a condition of receptivity.[9] More than this, it is absolute and singular—an appropriation and vindication, in fact, of Newtonian absolute space.

This leads us to the main point. Cassirer's categories—including space and time are all *active principles of functional unity, i.e., ways of organizing experience*. But Kant's forms of intuition and categories are constitutive of experience in a much more complete way; they have no possible application beyond it. Cassirer's categories, however, as principles of organization must themselves be directed to that which is itself *already* spatial, temporal, and causal in some primordial sense. If this is the case, however, the a priori status of his categorial field has to be explained in epistemological terms on a different basis than Kant's. Cassirer suggests that:

> The question of what, apart from spiritual functions, constitutes absolute reality, the question of the "thing in itself" may be in *this* sense, remains unanswered, except that more and more we learn to recognize it as a fallacy in formation, an intellectual phantasm. The true concept of reality cannot be squeezed into the form of abstract being; it opens out in to the diversity and richness of the forms of spiritual *life*—but of a spiritual life which bears the stamp inner necessity and hence of objectivity.[10]

Cassirer does not wish to deal with reality as it might exist independent of the mind's organization of it, for this would be too close to Kant's notion of the thing-in-itself—or, at best, too abstract a notion to be of any philosophical use. The closest which he comes to giving a description of it is in Volume 4 of *The Philosophy of Symbolic Forms* where he described it as an 'undifferentiated and infinite primordial ground'.[11] In Cassirer's terms, we can go no further. As he says,

> not only our world of concepts but also our whole world of perception are subject to characteristic conditions of form and are inseparably woven with them. But this positive context and cohesion also have their negative side. If we seek even hypothetically, to abstract from their conditions, then we have entered into a world that is to be conceived differently from ours not only as regards its form but also as regards its "matter". The reality of this world, the kind of phenomenal "givens" in it, cannot in any way be equated with our "world of experience". Of course we can never hope to make ourselves fully aware of this other reality in *conreto*, or even imagine it in all of its individual features. . . . Every such attempt at illustrative clarification must inevitably mislead us.[12]

Cassirer does not simply state this dogmatically, but actually shows at length the way in which vitalist philosophies (such as Simmel's, Klages', and Bergson's) that attempt to characterize some primal life-force are consistently forced into error. Cassirer also considers the view that a comprehension of the experimental world of other more basic organisms might yield the 'original metal' of reality before it undergoes 'minting' in human thought. This, however, 'provides us only with the picture's frame, never with its contents. It gives us nothing other than the physical outline of the gestalt, the "organization", and by itself, using purely empirical means, it can provide us with nothing more'.[13] In another posthumously published text entitled 'Basis Phenomena',[14] Cassirer's title suggests that he might here be negotiating with a mind-independent reality, but again his strategy addresses only certain basic forms of spiritual activity—namely the anthropological constants of 'I', action, and 'work'.

Cassirer's commitment to Critical Idealism, therefore, means that he is not prepared to countenance any conception of reality as it might be independently of the forms given it by human consciousness. This creates difficulties. It not only leaves the epistemological basis of the a priori status of his categories unclear, but also makes other aspects of his position problematic.

For example, he holds that the symbolic forms are constantly developed and deepened in an on-going historical and logical progression. Hegel's conception of Spirit is likewise construed as a progression towards self-consciousness, but whereas for Hegel this progression achieves a resolution in total logical comprehension, for Cassirer it does not. It is an open-ended progression. We claim that if this is the case then we must construe reality as ultimately independent of the mind. It is the challenge of a mind-independent reality that makes our basic symbolic capacity evolve in order to deal with different directions of physical challenge that it must deal with in order to survive.

At several points Cassirer admits this point indirectly. For example, in *An Essay on Man* he says 'nature is inexhaustible, it will always pose for us new and unexpected problems.'[15] There are other ways in which the existence of a mind-independent reality impinges on Cassirer's philosophy in positive ways. For example, in *Determinism and Indeterminism in Modern Physics* he allows that 'the phenomena of nature are not such as to elude or withstand in principle the possibility of being ordered'.[16] This means that nature must itself be in some way *amenable* to being given form through symbolic activity. Such activity is not arbitrary in how it orders the world. Therefore, we must assume some primordial regularity as a condition of our symbolizing activities.

This fact comes to the fore whenever Cassirer actually describes the first emergence of the natural symbol. For example, he says that:

> On close scrutiny perception is a process of selection and differentiation which consciousness applies to the chaotic mass of "impressions". Out of the mass of impressions which pour in our consciousness in any given moment of time certain traits must be retained as recurrent and "typical" as opposed to others which are merely accidental and transient; certain factors must be stressed and others excluded as non-essential.[17]

We must ask what are these impressions of? What are our criteria of 'recurrent' and 'typical'? The answer can be nothing other than a spatiotemporal realm of causally interacting particulars whose existence is ontologically independent of the mind and its symbolizing activity.

This enables us to solve one problem and to open up another. The solution lies in the fact that if the human subject is bound up with a mind-independent world of spatiotemporal causally interacting particulars then in order to emerge and survive in that world, it must *necessarily* find ways of giving it a unified articulation.[18] Human self-consciousness could not evolve without space, time and cause as principles of unification, because the reality which it faces is itself spatial, temporal and causal. The status of thing, number and magnitude is more complex. But it could be said that without them space, time and causality would not be symbolically articulable and thence could not function as elements in self-consciousness.

This gives a final answer to the a priori status of Cassirer's categorial field. It is a logical precondition of experience in the sense that we could not have experience of an objective world without it. But whereas in Kant such a field is no more than this, in Cassirer's theory it must also be seen as an evolutionary response to the demands of human inherence in a world that is itself of a spatial, temporal, and causal character. Its brute existence is mind-independent, even if its objective structure (as experienced) is not.

And this leads to the further problem. J.M. Krois has rightly observed that Cassirer is not a philosopher who merely proposes the primacy of consciousness, but declares rather 'The primacy of meaning and the fact that the person is bound to a body'.[19] Unfortunately, the relationship between the two factors identified by Krois is not well established by Cassirer. We have just shown how his categories are a priori in an evolutionary sense but his account of how the symbolic function actually emerges from embodiment is more elusive.

However, if we look at the *phenomenological* aspect of Cassirer's main operating concepts in more detail, then the nature of this emergence becomes clearer.[20] Cassirer's project relates to phenomenology in two different ways. First, there is an obvious parallel between it and Hegel's *Phenomenology of the Spirit*. As shown earlier, both Cassirer and Hegel see the formative activity of consciousness as not given once and for all but as logically progressing to more advanced degrees of comprehension. In relation to Hegel's philosophy, Cassirer says that for Hegel this comprehension issues in the primacy of the purely logical dimension of spiritual activity.

> Indeed, this ultimate reduction of all cultural forms to the *one* form of logic seems to be implied by the concept of philosophy itself and particularly by the fundamental principle of philosophical idealism. For if we renounce this unity, a strict *systematic* understanding of these forms would seem to be unattainable.[21]

The only alternative to this would be a strictly empirical approach, which merely describes the specific scope and content of different symbolic forms, as it were side by side with one another. However, this yields a disturbing dilemma.

> If we hold fast to the postulate of logical unity, the universality of the logical form threatens ultimately to efface the individuality of each special province and the specificity of its principle—but if we immerse ourselves in this individuality and persevere in our examination of it, we run the risk of losing ourselves and of finding no way back to the universal.[22]

This dilemma is, however, resolvable. As Cassirer says,

> If we can find a medium through which all the configurations effected in the separate branches of cultural life must pass, but which nevertheless retains its particular nature, its specific character—we shall have found the necessary intermediary link for an inquiry which will accomplish for the *totality* of cultural forms what the transcendental critique has done for pure *cognition*.[23]

The resolution posited here is, of course, the one proposed by Cassirer himself. The unity of the spirit in its logical and historical diversity does not demand Hegel's absolute idealism, but rather an identification of the individual formative principle, which is common to all this diversity.

In order to really underline his difference from Hegel, however, Cassirer needs to underline the way in which his philosophy goes beyond Idealism *per se*. This demands phenomenology in a second sense, namely one, which describes how the symbolic function emerges from the unity of the body itself. Fortunately, Cassirer does provide a great deal of material, which leads in exactly this direction. He effectively states the general question as follows.

> The relation between body and soul represents the prototype and model for a purely symbolic relation, which cannot be converted either into a relation between things or into a causal relation. Here there is originally neither an inside and outside nor a before and after, neither an agent or an effect; here we have a combination which does not have to be composed of separate elements but which is in a primary sense a meaningful whole which interprets itself, which separates into a duality of factors in order to interpret itself in them.[24]

What Cassirer is emphasizing here is how the unity of mind and body comes to know itself through the symbolic function. Cassirer takes us towards what Crowther calls *corporeal phenomenology*. It is well summarized in Cassirer's remarks that 'the hands are so closely bound up with the intellect that they seem to form a part of it'.[25] This line of thought probes to a level, which is deeper than Cassirer's 'official' description of his philosophy as a mode of Critical Idealism. The basic symbolic function is *the* constructive principle of experience, but it is more than just an act of mind which (*a la* Kant) gives form to experience. Rather it emerges from a more primal unity of the body's gradual co-ordination in response to the demands of its inherence in a primordial realm of spatiotemporal causally interacting particulars. As Cassirer states:

> At first local distinctions are closely linked with specific material distinctions—and it is eminently the differentiation of the parts of his own body that serves man as a basis for all other spatial specifications. Once he had formed a distinct representation of his own body, once he has apprehended it as a self-enclosed and intrinsically articulated organism, it becomes, as it were, a model according to which he constructs the world as a whole In this perception of his body, he possesses an original set of co-ordinates, to which in the course of development he continually returns and refers—and from which accordingly he draws the terms which serve to designate this development.[26]

To see the symbolic relation as itself emergent from this level may possibly have struck Cassirer as detracting from its autonomy. (The avoidance of this impression may well be why, he chooses to affirm his position as an Idealism, instead of a corporeal form of phenomenology). However, there is in fact no reason why its origins in embodiment should actually detract from the symbolic function's autonomy. Emergent and tertiary properties of things can often be analyzed as functions of more basic elements. But as Cassirer himself shows throughout his philosophy the functional unity of a whole is more than a mere sum or mere aggregate of its parts. If this is so then the autonomy of the symbolic function is not impaired by its origins, in the unity of the body, rather it is simply more complex than Cassirer's Idealist characterization of it.

The key point is that Cassirer's enterprise needs to be emphatically recontextualized in terms other than his official Critical Idealism. Kant's method gives him a decisive clue and a working methodology, but Cassirer develops it (without emphasizing the fact) in the direction of a corporeal phenomenology. This phenomenology is one, which gives primacy to the emergence of meaning from embodiment.

It is clear that this gives Cassirer's work much affinity with that of Merleau-Ponty. Merleau-Ponty, for example, notes that:

> It is no mere coincidence that the rational being is also the one that holds himself upright or has a thumb which can be brought opposite to the fingers; the same manner of being is evident in both aspects.[27]

This quotation reveals how, like Cassirer, Merleau-Ponty sees a necessary connection between human intelligence and the physical mode of its embodiment. Cognition—be it intellectual or our practical mode of engaging with the world—is not a purely mental operation, but a complex synthesis which draws on all our sensory and motor functions, as well as factors of memory and anticipation. These all link together to form a unified *field* of cognitive operations.

It might be possible to develop many analogies between Cassirer and Merleau-Ponty, but there is one important disanalogy. Our basic corporeal orientation is for Merleau-Ponty primarily a *pre-reflective* phenomenon. To deal with and find our way through the world does not require continuous conscious direction. Indeed, for Merleau-Ponty the great vice of empiricist and rationalist traditions is that they try to reconstruct the basic mode of our inherence in the world on the basis of models drawn from the domain of reflective thought—of which science is the best example.

As Merleau-Ponty says,

> To return to things themselves is to return to that world which precedes knowledge, of which knowledge always *speaks* and in relation to which every scientific schematisation is an abstract and derivative sign-language, as in geography in relation to the countryside in which we have learnt beforehand what a forest, a prairie or a river is.[28]

Merleau-Ponty is right to posit a corporeal orientation which precedes higher-level symbolic articulations of it. He emphasizes the direction in which Cassirer's own analyses tend, but which Cassirer himself does not underline properly. However, Merleau-Ponty suffers from the opposite vice. True enough someone can find their way through a landscape without knowing geography, but geography and other analytic disciplines are only particular employments of a general capacity for reflection and analysis.[29]

This capacity is a necessary condition of embodied *subjectivity* as opposed to mere animal corporeality. A subject would not be self-conscious unless, if called upon, it could describe who it was, where it is and what time it existed in, and other facts of an *explicitly* reflective nature. Furthermore areas of knowledge such as science, philosophy, history and art all embody high degrees of reflective awareness. Most of our cognitive life is pre-reflective, but more reflective modes of knowledge—concerning facts, figures, positions, and the like—often tacitly inform our pre-reflective orientation. If, therefore, a comprehensive corporeal phenomenology of the human condition is sought for, it must be one that can trace the structure of both the pre-reflective *and* reflective axes of consciousness *and* the principle of continuity which allows the latter to emerge from the former. This is precisely what Cassirer's philosophy of symbolic forms is able to provide. Merleau-Ponty is in a sense, locked in a notion of 'being-in-the-world' (i.e., basic corporeal orientation) in such a way that his capacity to deal with the higher symbolic orders of experience is restricted. (We will return to this question of the reflective standpoint in Chapter 2.)

III Symbolic Form and the Significance of Myth

The basic symbolic function, as we have seen, involves recognizing a particular as representative of a greater whole. This basic structure is embodied in different ways under different circumstances. There are varieties of symbolic form. One supposes that, for Cassirer, any distinctive way of using a symbol system, be it a particular mode of language-use, or system of notation or codification, or use of a particular medium—such as painting or sculpture, will be enough to count as a particular symbolic form. However, there are, for Cassirer, more general symbolic forms that he regards as fundamental. He declares that language, myth, scientific knowledge, and art are the most important ones—the basic cognitive modes that let us make sense of, and operate on the world. In *An Essay on Man*, history is added to this list.

Presumably there are different varieties of symbolic form because the basic categorial framework—of space, time, causality, thing, number and magnitude—operates with different emphases in different contexts of existence. History, for example, is language that gives special attention to causal events viewed in the horizon of time; mathematics and the sciences offer analyses and operative access to phenomena through various modes of quantification and observation, whilst the arts investigate structural features of the world through the sensible and/or imaginative embodiment of subjective experiences of it.

In Cassirer's philosophy, symbolic forms are not static. They achieve progressive articulation. He tells us that

> it is not a matter of disclosing the ultimate, absolute elements of reality, in the contemplation of which thought may rest as it were, but of a never-ending process through which the relatively necessary takes the place of the relatively accidental and the relatively invariable that of the relatively variable We can never claim that this process has attained to the ultimate invariants of experience, . . . Rather, the possibility must always be held open that a new synthesis will instate itself and that the universal constants, in terms of which we have signalized the "nature" of certain large realms of physical objects, will come closer together and prove themselves to be special cases of an overarching lawfulness. It is this new law that will then form the true nucleus of objectivity; but it, too, must expect to be recognized in its merely contingent universality and replaced by a still broader universal relation.[30]

Cassirer also makes it clear that each form has its own logical principles, and that individual articulations within it are always made meaningful by their position within the developing (as it were) grammar of that form, and the way it is used. Features that, at one stage of development, seem essential, are superseded by new ones that make the form more unified and comprehensive in its explanatory value. Equally importantly, progression within a specific symbolic form will always be affected by its relation to the evolving totality of symbolic forms.

We reach a potential difficulty. The remarks by Cassirer quoted immediately above aptly emphasize the open-ended and constantly developing aspect of symbolic forms as such. However, the content of his remarks clearly fits best those symbolic forms which look to comprehend the world in terms of its general features and laws of behaviour. But Cassirer himself also warns us that, 'High as myth and art may rise, they remain on the soil of primary, 'primitive' expressive experiences.'[31] This does not sit easily with the remarks last quoted. Neither does historical understanding, as it is very often concerned with comprehension of the concrete particular in terms of lives and their circumstances, rather than with extracting some general truths from them.

In the case of cultural phenomena, especially art and aesthetics, then, if the concept of progressive articulation is to have application, it must be carefully qualified. Cassirer provides an important guiding principle in relation to this. In relation to the development of a symbolic form, he says that:

> This operation has essentially a *critical* character. Elements hitherto accepted as certain, as objective, are continuously rejected when it turns out that they do not fully accord with the unity of experience, or at least that, measured by this unity, they possess only a relative and limited and not absolute significance. It is at all times the order, the necessity, of phenomena *as a* whole that serves as a criterion for the truth of the *particular* empirical phenomenon and of the 'reality' that should be imputed to it.[32]

Here, the important point is that the form develops through its elements having a developing critical relation to one another within the context of the form as a whole. In the sciences this may involve increasing unity and comprehensiveness of explanatory scope, but Cassirer's approach clearly allows that within other forms, progression can be achieved by the emergence of new elements. These move things forward by reconfiguring the overall field by means of their critical relation to other elements.

All this being said, Cassirer's corporeal phenomenology is still incomplete. The problem is that he does not have an account of the relation between the reflective standpoint and the unity of the self. Clearly the symbolic function in itself offers a basic structure—because we know ourselves as a unity where the individual moment of awareness expresses a more general personal history on which it depends, but to whose emerging character it contributes. But this needs to be described in more detail because, for

Cassirer, our knowledge of the unity of an objective world is correlated with knowledge of the self. As he puts it,

> 'inner and outer experience are not two foreign and separated things but rather . . . they are based upon common conditions and . . . they can constitute themselves together and in constant reference to one another.'[33]

The reason why experience of objects is correlated with spiritual activity is because the objective world is not simply there. It is so in an overwhelming variety of different aspects. To make sense of these aspects—to know it *as* objective, we must approach it selectively on the basis of features that have engaged our attention and interests. And likewise, this only happens because reality's multifarious aspects are amenable to being acted upon—they allow spirit to express itself, and come to know itself through such activity.

Art and the aesthetic have a special involvement in this. They define the way in which a cultural group objectifies the world through finding a home and identity in it at the sensible level. As S. G. Lofts says of Cassirer's position:

> Each culture gives itself form and a sensuous presence through its aesthetic comportment towards the world. Not only architecture, language, dress, gestures, and mannerism, but also the body and face are formed and sculpted according to a common aesthetic norm that defines the identity of a culture and time. This cultural aesthetic appears in the sensuous presence of reality itself.[34]

Closely related to art and the aesthetic (as a symbolic form where the reciprocal interaction of objective knowledge and subjective awareness is shown very emphatically) is mythic consciousness. The emphasis that Cassirer gives to this as a symbolic form is one of the most unique features of his thought. For him, the basic symbolic function finds its most primal embodiments in language, art, and mythic thinking. In mythic thinking, the distinction between the world as object and the attributes of consciousness itself has not yet been clearly made. It is closely correlated with physiognomic properties, and the expressive levels of human consciousness. In Cassirer's words.

> The world of myth is a dramatic world—a world of actions and forces, of conflicting powers. In every phenomenon of nature it sees the collision of these powers. Mythical perception is always impregnated with these emotional qualities. Whatever is seen or felt is surrounded by a special atmosphere—an atmosphere of joy or grief, of anguish, of excitement, of exultation or depression. Here we cannot speak of "things" as a dead or indifferent stuff. All objects are benignant or malignant, friendly or inimical, familiar or uncanny, alluring and fascinating or repellent and threatening.[35]

This is the 'sympathetic' orientation of myth. Mythic consciousness regards itself as totally inherent in an emotionally and affectively significant realm of living forces. Such a realm is conceived, in effect, as a 'society' of such forces. The mythic thinker has the sense of being an aspect of a '*solidarity of life*' that encompasses and absorbs the great multiplicity of different individual life forms. Such a thinker does not regard the human being as having some especially privileged place in all this. All forms of life inhabit nature in basically similar terms.[36]

As might be expected, Cassirer takes the conceptual framework of mythic thought to be an extremely loose one:

> all the boundary lines drawn by our empirical concepts of genera and species keep shifting and vanishing. Not only does one and the same being perpetually take on new forms; at one and the same moment of its a existence it also contains and combines within itself an abundance of different and even mutually opposed forms of being.[37]

The mythic mind does not think in terms of the concept of things or concepts of causal relation but rather personifies these, as expressions of divine or demonic will, and the like. Of course, as the human species develops, cognitively clearer formulations of the symbolic function and ways of logically developing it are gradually realized, and we find a differentiation of mythic consciousness into the various modes of symbolic form.

However, we will also recall Cassirer's observation that 'High as myth and art may rise, they remain on the soil of primary, 'primitive' expressive experiences.' Mythic consciousness may be articulated into other symbolic forms, but there is a sense in which it remains a part of us, and is connected to us having aesthetic and artistic needs. Cassirer does not explain what this involves in any detail. However, it must surely have something to do with the particular way in which self-consciousness embodies the basic symbolic principle.

What complicates this is not only self-consciousness's deep link to the affects, but surely also the fact that its unity is of a more open kind. Indeed, some other remarks by Cassirer concerning mythic consciousness offer some important insights on this. He suggests that:

> Because myth lacks the form of causal analysis, it cannot know the sharp dividing line which only this form of thought creates between the whole and its parts. Even where empirical intuition seems, of itself so to speak, to give us inwardly differentiated things, myth replaces this sensuous separation and contiguity by a characteristic form of interpenetration. The whole and its parts are interwoven, their destinies are linked, as it were—and so they remain even after they have been detached from one another in pure fact.[38]

These remarks open the following line of argument. Mythic consciousness involves no clear separation between the self and the world in terms of objective concepts. As the basic symbolic function is further articulated, however, a clear logical distinction between the world of objects and events and that of our inner life is gradually established. But vis-à-vis the unity of self-consciousness, the relation between individual moments and the developing whole of an individual's experience remains a very close one. They involve a *holistic* interpenetration—each is a necessary part of the definition of the other. The relation between individual experiences and the whole, in other words, retains something of the intimacy that characterizes how mythic consciousness imagines its relation to nature.

However, it is vital to emphasize that this link to the mythic is not some cognitive regression that we are unable to overcome. In fact, the clear distinction between the objective world and the subjective experience, as well as the notion of experience itself, actually presupposes the capacity for *imagination*. As we shall see later, this has an extraordinary effect in terms of how we think of both the objective world and of the unity of our experience, and the two together. Imagination represents in sensuous rather than conceptual terms, and it is this sensuousness of the imagined—with its frequently strong affective charge—that gives our sense of self in relation to the world a continuing residual affinity with mythic consciousness. It is the deepest source, also, of our need for aesthetic and artistic expression, and their significance as symbolic form. (The forgetting of this by contemporary aesthetics is, arguably, the main reason for the discipline's marginalization.)

Now, as Donald P. Verene has shrewdly noted, Cassirer himself has no well-developed theory of imagination, and relies very much on Kant's approach (in particular the Schematism of the Categories).[39] Indeed, beyond his citations of the authority of Kant, he does not explain the role of imagination in objective knowledge and knowledge of the self, in any detail.

However, Crowther has offered such explanations in depth. Indeed, he takes them in a further direction that is of particular use to our book. Recent commentators have linked the unity of the self to narrative structure; Crowther does so as well, but argues that narrative unity involves aesthetic pleasure and that the aesthetic is central to both its ontogenetic emergence and later consolidation. His way of showing this not only bolsters Cassirer's corporeal phenomenology, it offers further material for understanding the origins of the aesthetic and artistic as symbolic form.

Notes

1 *PSF*, Vol. 1, p. 77.
2 *PSF*, Vol. 3, p. 127.
3 *PSF*, Vol. 1, p. 78.
4 Ibid., p. 79.
5 Cassirer actually makes this point about art in *Essay on Man*, p. 168.
6 *Substance and Function*, p. 269.
7 *PSF*, Vol. 4, p. 21.

8 Immanuel Kant, *The Critique of Pure Reason* trans. N. Kemp-Smith (London: Macmillan, 1989), p. 149.
9 See, for example, Immanuel Kant, *The Critique of Pure Reason*, trans. N. Kemp-Smith (London: Macmillan, 1989). On p. 69 Kant states unequivocally that 'space is not a discursive or, as we say, general concept of relations of things in general, but a pure intuition'.
10 *PSF*, Vol. 1, p. 111.
11 *PSF*, Vol. 4, p. 31.
12 *PSF*, Vol. 4, pp. 54–55.
13 *PSF*, Vol. 4, p. 52.
14 Included in *PSF*, Vol. 4.
15 *EM*, p. 218.
16 In *Determinism and Indeterminism in Modern Physics* (New Haven: Yale University Press, 1956), 60.
17 *PSF*, Vol. 2, p. 35.
18 This suggests that the 'organic' experience of space and time which Cassirer identifies in *EM* (p. 42) remains an immanent factor even in rational experience.
19 John M. Krois, *Cassirer: Symbolic Forms and History* (London and New Haven: Yale University Press, 1987), 7.
20 Cassirer had originally considered entitling The Philosophy of Symbolic Forms by the title, which he eventually confined to the third volume—*The Phenomenology of Knowledge*.
21 *PSF*, Vol. 1, p. 84.
22 Ibid., p. 84.
23 *PSF*, Vol. 1, p. 84.
24 *PSF*, Vol. 3, p. 100.
25 *PSF*, Vol. 1, p. 184. The term 'corporeal phenomenology' is used by Crowther in relation to Cassirer in *Philosophy After Postmodernism: Civilized Values and the Scope of Knowledge* (London: Routledge, 2003), 44.
26 *PSF*, Vol. 1, pp. 206–207.
27 Maurice Merleau-Ponty, *The Primacy of Perception*, ed. James E. Edie, trans. Carleton Dallery (Evanston, IL: Northwestern University Press, 1964), 197.
28 Ibid., p. ix.
29 Crowther has offered a trenchant critique of Merleau-Ponty on these same lines in a paper entitled "The Poetry of 'Flesh' or the Reality of Perception? Merleau-Ponty's Fundamental Error", *International Journal of Philosophical Studies* 23, no. 2 (2015): 255–278.
30 *PSF*, Vol. 3, pp. 475–476.
31 Ibid., p. 449. As we shall see later on in this book, art can be interpreted as a symbolic activity, which admits of progressive articulation.
32 *PSF*, Vol. 2, p. 31.
33 *LCS*, p. 30.
34 Lofts, *Ernst Cassirer*, p. 185.
35 *EM*, p. 767.
36 Ibid., p. 82.
37 *PSF*, Vol. 3, p. 61.
38 *PSF*, Vol. 2, p. 51.
39 Donald P. Verene, "Cassirer on Culture", *International Philosophical Quarterly* 22 (1982): 133–144.

2 Crowther, Imagination, and the Origins of Symbolic Form

As we have already seen, for Cassirer art and the aesthetic are expressions of a symbolic form in which the physicality of the real is selectively ordered on the basis of spiritual activity. Through this, we make ourselves at home in the objective world. For Cassirer, imagination is central to this. Paul Crowther holds the same position and presents it in a way that is more amenable to developing the scope of aesthetic cognition than is Cassirer's own source—Kant. In this chapter, then, we will consider Crowther's arguments concerning the extraordinary scope of imagination in permeating all aspects of our existence. It is this ubiquity that explains why aesthetic experience and the arts are pursued and valued so highly.

I The Necessary Role of Imagination in Objective Knowledge, and Knowledge of the Self

According to Crowther imagination is the main feature that correlates our knowledge of an independently existing 'objective' world, with the unity of self-consciousness.[1] Of course, 'imagination' is a term used in many different senses but Crowther emphasizes one in particular—imagination interpreted as *the capacity to generate mental imagery with a quasi-sensory character*.[2] For Crowther, imagination understood in this sense is the basis of the '*trans-ostensive*' factor in cognition—the feature that allows it to represent things which are not immediately present to perception.

Trans-ostensive reach is basic to all knowledge. This can be shown, first, by following Crowther's explanation of its role in knowledge of objects. In this, he makes use of an important passage from Gareth Evans.

> The capacity to think of oneself as located in space, and tracing a continuous path through it, is necessarily involved in the capacity to conceive the phenomena one encounters as independent of one's perception of them—to conceive the world as something one "comes across". It follows that the capacity for at least some primitive self-ascriptions—self-ascriptions of position, orientation, and change of position and

> orientation—and, hence, the conception of oneself as one object amongst others, occupying one place amongst others, are interdependent with thought about the objective world itself.[3]

According to Crowther, Evans's remarks clarify that to recognize the world as existing independently of one's perceptions; one must be able to chart one's course through it—with reference to one's past positions and possible future ones. We become aware of ourselves by reference to the things we have encountered. At the same time, to recognize past, present, and possible positions of one's embodied self is to become conscious of a world whose aspects one 'comes across'—that endure independently of us, and are there to be encountered again when we move into the right position. Self-recognition and object recognition are correlated by reference to the relation between what is given, and our awareness of ourselves as present, and reference to other possible positions that are not immediately given.

It might seem that it is language that provides the means for trans-ostensive reference. However, whilst language is vitally connected to this in the mature subject, Crowther argues that imagination is equally important and is actually presupposed in order to explain the acquisition of language itself. It provides the means whereby signs and symbol get trans-ostensive reach. Our learning of signs of whatever sort at the crudest cognitive level requires that we become habituated to using them ostensively, but have some means of representing possible applications of the sign to things that are not immediately present in perception.

The claim that imagination is a transcendental condition of language and knowledge links Crowther to Kant and Cassirer, but separates him from Edward S. Casey, J-P. Sartre, and Mark Johnson, who have all offered important discussions of the phenomenology of the image but have not assigned it a necessary role in the relation between objective knowledge and knowledge of self.[4] Colin McGinn's work links to Crowther's position by arguing that that imagination is a prerequisite for knowledge through its role in language acquisition.[5] However, he does not offer a detailed account of its ontogeny. Crowther does so, on the following broad lines.

In animal consciousness, a creature recognizes things through habituation to them or through instinct. But this is not to recognize through concepts. One has not understood the concept of something unless one knows that the concept term is applicable to more than one example of the relevant class, and that it applies to members of the class encountered in different times and places. To learn even the crudest sign (let alone complex concepts) requires that we are able to follow rules, but this logically presupposes the involvement of trans-ostensive considerations. We cannot learn a term without knowing that it can relate to individuals other than those immediately present to perception, and without knowing how to apply it.

Now if this trans-ostensive dimension was itself explicable as a special kind of rule following we would be led into an infinite regress of rules explaining rules. To avoid this, we have to assume some non-linguistic competence as the basis of our trans-ostensive orientation. Imagination is the best candidate for this. Put generally, imagination is the component of cognition that allows the human subject to represent *what it might be like* for he or she or other objects to be elsewhere in space and/or time. Through exercising this capacity, our perception of the world has, as it were, no blank spaces. Even if we can't perceive some object, state of affairs, or situation (or aspects of them) because of their hiddenness, or present inaccessibility, we can project in imagination what they might be like.

Suppose, for example, that there is a building in the distance. Some of its sides are hidden from us, and we cannot see the rooms inside. However, we can, at least imagine what these hidden or currently inaccessible aspects might be like. The key point is that to imagine something is to generate an image of it that has general phenomenal consistency with those features of the object that we do know through perceptual acquaintance. Enough features are projected to be consistent with how that individual *might* appear if we could directly perceive it.

For Crowther, then, imagination allows us to form a sense of the world's spatiotemporal continuity beyond what is immediately given in the perceptual field. As well as being essential to concept-formation, this is also essential to the human power of attention. For we have a sense of something enduring through time in immediate perception, only insofar as we can project how the object appeared a few moments earlier and anticipate how it might appear a few moments thence. Imagination gives orientation to our power of attention.

It should be emphasized that, in Crowther's terms, whilst imagination is a non-linguistic capacity that enables the formation of concepts and language, as these develop, the exercise of imagination comes to be very closely guided by them. Of course, one's imagination can wander, but this is not the point. Once language has been acquired it allows imagination to be directed and exercised in a fully volitional way. If however, imagination was not already in position—as a cognitive capacity for the trans-ostensive extension of the immediately given—then the very idea of learning would be unintelligible.

It is worth dwelling further on the important role that Crowther assigns to imagination in the ontogeny of perception, because it is here that link to the basic symbolic function first emerges. For him, perception centers on a unified sensorimotor orientation towards the overall field. Our intentional activity—the directedness of consciousness—means that in our relation to objects or to states of our own being, some sensory factors will be attended to in closer terms than others. For Crowther, this important because the infant's initial learning of the unity of objects and physical states of affairs is not just a case of recognition. The child learns the basic symbolic function (the way a whole is implied by its parts or aspects) through practical exploration. By exploring and playing with the object through physical efforts of grasping and the like, the infant gets a sense of what can and cannot be done to the object, and which physical conjunctions it is amenable to, and which it is not. In this way, the infant gradually learns the whole/part or aspect relation as a general style of being. Crowther uses the term 'style' here because the unity of the object not only involves a particular physical mode of world occupancy, but also positions the observer in terms of awareness and/anticipations of what it allows and disallows vis-à-vis perceiving it and physically manipulating it.

Crowther emphasizes that in this style of being, there is no separation between particularity and the possibility of generality. If a child kicks a ball it has not seen before, it has recognized a possibility of kickability, but has not yet articulated it in these general terms. As an infant explores its physical environment, these accumulating episodes form a whole which is more than the sum of the parts. The infant achieves a progressively unified coordination and control of its sensorimotor activities in relation to the demands of the physical environment. The individual explorations and episodes that the infant experiences come together in a general adaptation to three-dimensional space, time, causality, and multi-sensory phenomenal objects and configurations. In the course of this, the infant comprehends both individual and general patterns of phenomenal unity without, at first, distinguishing between them.

According to Crowther, the mimetic aspect of imagination—its representation of quasi-sensory aspects of that which it is an image of—is vital to all aspects of ontogeny. We cannot recall all the countless traces of rooms, dogs, showers of rain, etc. which we have perceived, but we do know what such things *are like*. This means that we can generate images or image-sequences which embody quasi-sensory properties that are generally isomorphic with those things or kinds of thing which they are images of. In this way, the imagination continues the recognition of individual and general patterns of phenomenal unity that are first understood through the body's practical modes of adapting itself to, and manipulating them.

The emergence of images in the infant's consciousness may at first be fragmentary in and tied to the experience of stimuli. However, as the infant's senses and actions become coordinated the generation of imagery becomes more systematic and controlled. When, say, handling an object, it will be able to project how those aspects of an object or place presently hidden from view might appear. By representing, in imagination, what these non-immediate aspects might be like, it comes to learn that these other aspects also belong to the object. This means that the objectness of the object has been fully appropriated at the level of the phenomenal. Imagination—even at the infant stage—extends the scope of the body's operational vectors by allowing non-immediate quasi-perceptual viewpoints to augment those which are available upon the immediately given. It opens up the trans-ostensive dimension.

It is the reciprocal interaction of this capacity with the infant being instructed on how to communicate through sounds, and pointing, and the like, that allows concept-acquisition and initiation into language. Crowther considers the example of a child being taught the concept 'dog' through the word being uttered whenever a relevant animal appears in view. In such cases, the child learns not merely in recognitional terms. It surveys dogs from different angles and viewpoints, and through touching and smelling as well as through seeing. In the case of these explorations it becomes habituated to characteristics of dog-appearance, and is able to form—however schematically—an image of phenomenal unity, of features that evoke dogginess.

The fact that the term 'dog' engages with both percept and image in the infant's consciousness, is further interpreted by Crowther as follows:

> The term 'dog' applied to both percept and image not only emphasises their shared logical object, but also—since the word can be used without the percept being present—allows the image to function as a progenitor of conceptual meaning in its own right. On hearing the word, for example, a child might say 'bow wow' or stick its tongue out and utter canine panting noises. A response on these lines indicates that the child can recognise what a dog is, and, more significantly that it has formed an image of general canine behavioural traits. This latter point provides evidence that the child is beginning to understand 'dog' as a term which ranges across different individuals, times and places. It has acquired some bare rudiments of 'dog' as a concept.[6]

Crowther's point is that the trans-ostensive fecundity of the image qua image is reciprocally correlated with the acquisition of language and concepts. It allows referring terms to be learned as having extension across non-immediate contexts, just as the learning of those terms allows it to be directed volitionally—become subject to the will. Once a basic vocabulary involving kind and relational concepts, and proper names, has been acquired, then we already have a rudimentary basis for knowledge of the world as objective—as an order of being which exists independently of us.

II The Phenomenology of Imagination

As well as investigating the cognitive function of imagination Crowther is equally attentive to the phenomenology of the mental image itself. This is the basis of his explanation of imagination's major role in the unity of self-consciousness. It is interesting that Cassirer says that:

> In man we cannot describe recollection as a simple return of an event, as a faint image or copy of former impressions. It is not a repetition but

> rather a rebirth of the past; it implies a creative and constructive process. It is not enough to pick up isolated data of our past experience; we must really *re-collect* them, we must organize and synthesize them, and assemble them into a focus of thought.[7]

Crowther argues that this 're-collective' aspect of memory is due to the role of imagination, and the intrinsic creativity of image generation. He identifies a number of things that are basic to this creativity. The most important of them is its schematic structure. Because imagination is subject to the will, it does not mechanically reproduce that which it is an image of, it generates it very selectively—omitting some details, emphasizing others. Indeed, the image, *qua* schematic, interprets its object with considerable licence. In imagining say, a dog, we may simply project a creature with fur, a tail, and two floppy ears. It need not be exactly like any dog we have actually seen. And even when we do imagine a familiar individual, we do not reproduce it in its entirety.

The fact that the image is schematic means that, in the normal mature human subject, imaginings can be clearly distinguished from perceptions. The distinction here is qualitative rather just one of degree. Whether or not an object can be perceived depends on it being physically available to perception. Imagining an object is not so dependent, and the fact that (*qua* schematic) it is an interpretation of its object means that even the objective aspects of its appearance—shape, mass, volume, size, and position, and the like, are posited in a way that is distinctively ours. The image interprets the object in ways consistent with the features that define it as an individual member of some class, but presents the relevant from an individual viewpoint.

Under normal conditions, our imagining of something simply does not require that a particular formula for selecting individual aspects is followed. A preference may exist—as in everyday perception itself—for imagining the object through frontal, 'normal' sized aspects. But we are by no means tied to this. Our imagining an object can generates a schema of its aspects in any sequence, and from any angle desired. When imagining events, indeed, whilst the component factors must be represented in a linear temporal order, the character of the features that compose the event are very determined by the imaginer's personal sense of what that event involved, or, if it is merely an imagined possibility, what it might involve.

In this way, it represents not only objects, but, again, embodies a *style* of perceiving them. The reason for this is that, because the generation of images is subject to the will, how we project the object will derive from our unique personal history of perceptions and conceptions. In this way, the image always indicates its subjective origin. What we imagine is, in effect, experienced as if it were present to us. This is not just because it is we who have generated the image, but because how we project it is determined by our personal history in relation to the thing, or kind of thing, or situation, that we are imagining. In what follows, therefore, instead of talking of the image's schematic nature, we shall emphasize Crowther's characterization of it as a stylized, creative *interpretation* of its object.

This, we should emphasize, is of special importance in explaining the residual cognitive presence of mythic consciousness. That consciousness emphasizes the intimate bonding of self and nature. As Crowther puts it,

> In imagination, the appearance of what we imagine has a style determined by its relation to that accumulating whole of experience—the intentional nexus—which sustains our willing (a whole that is, in effect, our personal history). Style in the image means that *the object appears differently than it does in real life, and the difference is down to us*. . . . In this way the object is *inhabited* by us.[8]

The subjective dimension of imagination is also emphasized by another feature that Crowther emphasizes, namely its *radical instability*.[9] The image's details, and the vectors and unity of its appearance usually shift and reconfigure in some way even as we project them before the mind. In people who suffer from hallucinations and the like, it may be that images have—on occasion—a stability and intensity that makes them indistinguishable from real perceptions. But if there were not a gap between perception and imagination then the very notion of experience would become unintelligible. We would not be able to distinguish between the actual and the possible. (Of course, whilst imaginative projection has the affinity with mythic thought just noted, it is this feature that establishes a clear gap between the image, and independently existing objects of perception.)

Crowther emphasizes another important phenomenological feature of imagination—its *creative plasticity*. Imagination allows us to represent not only what an object's hidden or inaccessible aspects might be like, but also transform how the object appears—adding some features, taking others away, and subjecting it to transformations or unexpected conjunctions. We can even change it into something entirely different. Obviously, we cannot imagine things that are logical impossibilities—such as square circles, and we cannot imagine all the details of vast or phenomenally extremely complex objects, but apart from these limits the imagination has great freedom to project possible objects and affairs according to its will. This allows for the most unlikely images, such as talking rocks, flying elephants and the like.

There are some important weaknesses in Crowther's approach. The first is that his account is directed very much towards objects and events as such (especially visual ones). He devotes much less attention to the fact that imagery often occurs in associational sequences—often with a narrative element. The second is that he treats the image's radical instability as a matter of phenomenological fact rather than an as something with deeper existential meaning. We hold that whilst this instability allows a cognitive distinction between image and perception to be made, it nevertheless, involves some sense of loss—it is something that we would like to keep, although we know that it is only transient. A third problem is his failure to emphasize the importance of the involuntary generation of images in the life of self-consciousness. A final problem is that whilst Crowther identifies the key elements in imagination, he does not pay enough attention to the experiential whole which is emergent from these. His important discussions of ontogeny show how imagination comes to realize its transcendental function as a condition of knowledge and self-consciousness, but he overlooks the fact that this is, at the same, time an initiation into what might be called *the life of imagination*. How we exist in the world extends much further than the mere recognition of objects and our movements amongst them. We dwell *in* this relation. The things we perceive, and the memories we recall are profoundly influenced by the things we imagine in relation to them and by the imaginings they provoke involuntarily—daydreams, possible goals, things that might have happened in a different order of events, and the like. These are not incidental to our pursuit of practical projects, careers, and the search for instinctual gratifications, *they permeate them*.

Cassirer and other philosophers have emphasized the importance of emotions in the life of the subject. But emotional responses and attitudes are not blind. They are emergent from what happens to us, and just as much from how we imagine things to be. Imaginings wash over our everyday projects, stimulating strong or vague emotional responses and many shades of feeling between.

This is why Crowther should have addressed the meaning of imagination's radical instability, and its involuntary expressions, in more detail. These are important affirmations of the bitter/sweet intonations of our finitude. What we have, as finite beings, in perception, and life in general, is never possessed absolutely. As something is identified or attained, other things crowd in to demand our attention. What we desire is realized or thwarted, and then passes away into imaginings or rememberings of it. Indeed, our fondest images—be they employed in remembering or in fantasy, are not the kind of things which can be exactly formulated and made unchangeable. Just as the objective world changes as we move through it, and through factors beyond our control, so too, in correlation with this, do our desires, our cognitive orientations, and the imaginings that are at work in these.

Indeed, as well as the imaginings that arise in relation to immediate goal-orientated activity, a vast range of involuntary associations are generated alongside these—taking us to places other than that of the immediately given. The practical vectors of experience are embedded in a broader continuum of acts of imagination, and associational imagery (voluntary and involuntary) that never resolve into some definitively complete state of self-possession. Through this, the ontological frailty and transience of our inner-life becomes something felt.

III The Narrative and Aesthetic Unity of the Self

In later chapters we shall argue that the image-rich *act-association continuum* just described, is vital in establishing the recurrent importance of art. However, before we approach that question we must consider one other important feature of Crowther's work. Whilst he does not formulate the aforementioned act-association continuum as an overt operational concept, he is indirectly aware of one of its major consequences. The unity of self-consciousness is more than the capacity to ascribe experiences to oneself. It is determined by a narrative structure.

The role of narrative in personal identity has been a key philosophical concern since Alasdair MacIntyre's formulations of it, but Crowther's approach is unusual in arguing that the role of narrative in personal identity involves *aesthetic structure*.[10] That it does so is not surprising, given the act-association continuum that we have just described. For him, generally speaking, a narrative is unified set of descriptions centering on the actions and/or experiences of sentient creatures. In fact, unification is the actual purpose of narrative—it identifies a pattern of meaning that allows a series of events to be understood as owing much of their particular character to their role in constituting the meaning in question.

Many narratives are, of course, fictional, but for Crowther they do not have to be. Many narratives are true stories, not least of which is the personal history of an individual's life. The passage of our lives involves a succession of facts about what we did, and what happened to us. However, in order to act and perceive we must relate to our past and the realm of possibility in highly selective terms. Personal history has something of the character of imagination, and, of course, imagination is central to how we make sense of our personal history. This is because, in addressing our past, and projecting our future, we often do so through making it come alive in imaginative terms. Indeed, a great deal of our introspection involves imagining what would have been the case had things turned out differently in the past.

Crowther rightly emphasizes that this has a profoundly unifying effect through, in effect, making the past or future answer the needs of the present. As he puts it,

'Imagination allows us to possess the memory—to inhabit it as something living rather as a mere fact about our past'.[11] This is also true of how we imagine future or counterfactual possibilities. Our imaginings stylize their content—giving them emphases or making omissions that reflect the nature of our present interests and orientations. This makes the past, future, and counterfactual possibilities matter to us in emotional terms, over and above their importance as considerations that guide our practical planning. The creativity of imagination infuses all aspects of our personal history.

Of further importance is the fact that none of our actions or experiences can be described exhaustively through a single description. The things we perceive and experience are describable, in principle, under many different aspects. Similar considerations hold in the case of how we connect descriptions or imaginings of fact to one another, and to broader contexts of belief. All descriptive, imaginative, and discursive cognition is essentially selective, and is so on the basis of a number of factors. Crowther suggests that:

> Chief among these is relevance to our current position in life, and the things we value. On the basis of these we map out a route or series of connected routes through both our past, and the realms of futural and counterfactual possibility. The purpose of this mapping is to enable these elements to converge on the present, and thus make it intelligible as a stage in the ongoing unity of a particular life history. Self-consciousness coheres, accordingly, as a narrative structure.[12]

Elements of self-deceit may enter into this, and the same body of personal experience may be rendered intelligible in terms of different narratives, but these differences are by no means arbitrary—amounting to no more than a case of how one wants to tell the story. One might make sense of the passage of one's life in terms of pursuing a career or in terms of finding a mate and having a family life. In most circumstances, of course, our life-narratives will be variable and complex and relate to these and other considerations. As well as sometimes involving self-deceit, these narratives may even engender psychological conflict as our existential circumstances become more complicated. These conflicts can then be a focus of narratives themselves. However, Crowther emphasizes that

> no matter how diverse the narratives in terms of which we make sense of our lives *they must be logically consistent with one another*. (If this were not the case *chronic* schizophrenia would be the norm). The cohesion of self-consciousness, therefore, is constituted by a general narrative structure which is a function of the logical consistency of its component sub-narratives.[13]

The point is, then, that imagination is so infused in how we think about our personal history, and that history itself opens itself out only through various interpretative viewpoints, that the unity of self-consciousness must involve much more that personal history as some dry linear continuity. It involves a developing narrative that is emergent from a play of its component episodes. As we shall now see, this has an *aesthetic* dimension.

Again, Crowther links the origins of this creative narrative structure of experience to the ontogeny of perception itself. Much of an infant's cognitive activity is based on the satisfaction of basic biological needs or getting recognition from others. However, such a being also enjoys play, and the exploration and rearrangement of its physical circumstances for the sheer pleasure of it. We have already seen how Crowther discusses the importance of this in learning concepts. He also emphasizes that the play involved in the infant's cognitive exploration of its surroundings and engagements with other people is something that can be enjoyed for its own sake—it has an aesthetic character. Indeed, the evolutionary importance of aesthetic pleasure or displeasure is the way it stimulates such cognitive exploration. On the basis of such explorative aesthetic play, the infant gradually coordinates its bodily movements and is initiated into language. In an important passage, Crowther argues that this intrinsically pleasurable activity exceeds its perceptual function:

> It also has, simultaneously, an internal function, in facilitating the cohesion of the self. For in co-ordinating its body with the world, the child learns a basic framework of constants, possibilities, and impossibilities, which through articulation in language, is able to organise its powers of understanding and imagination. Specifically, it can form a sense of itself as a person with a history, not only through remembering things which it has done, but also by identifying with other people—either directly, or through play, or hearing stories. It only becomes aware of itself as a person insofar as it can selectively inhabit other personalities through imagination. This switching to and from between possible narratives allows it to focus and refine what is distinctive about its own personality. The narrative of self originates, in other words, in a playful rather than a rigid correlation of understanding and imagination. It has an aesthetic character.[14]

This aesthetic dimension is also found in adult experience. To define oneself involves more than a name and descriptions of our present life-situation. It centers on a complex play of aspects. These may relate to our past and present social roles; friendships; our involvement with a family or community; our religious beliefs (if any), and biological questions. Even monotonous or ascetic lifestyles involve an element of playful narrative in that, for example, the ascetic's withdrawal from life is given full definition only insofar as it plays off against imaginary involvements in life, is only made real insofar as it is defined, felt, or reaffirmed against a background of imagined involvement in life.

A sceptic might ask why any of this is connected to the aesthetic. Even if the unity of the self is built on interweaving narratives, there is no reason to describe this as a 'play'. Indeed, given that a widely accepted criterion of the aesthetic is its 'disinterested' character this can hardly have application to the self because the things we do, believe, and value, are surely the epitome of *interested* activity in the most uncompromising sense.

Crowther himself, addresses this problem, and links his theory to an interpretation and modification of Kant's account of the pure aesthetic judgment.[15] However, we prefer to take a different route by affirming the importance of what we earlier called the act-association continuum. Elements of creative modelling, play, and associational activity permeate all aspects of conscious life—from knowledge of objects as things that present different aspects to us, to our sense of who and what we as individual persons. To be an individual person is not just to continuously inhabit the same body, or to satisfy formal criteria for continuity of and self-ascription of experience. Experience matters to us because of the pleasure of having it, over and above the basic instinctual drive for self-preservation. The sense of who and what we are viewed in the context of the act-association continuum is unified, but its unity is an open, holistic one comprising elements that are themselves individually creative, and an emergent sense of self that gradually changes. The unity which comprehends this must have a relatively open and playful character—that is to say, a narrative.

It is precisely the importance assigned to imagination and narrative structure that allows Crowther's approach to highlight the centrality of the aesthetic in experience. The pervasiveness of the imagination in all aspects of self-consciousness means that a creative element is introduced simply through basic cognition. Indeed, our personal history and future considered as whole is more than the sum of the practical projects which are basic to them. Even with such concrete 'interested' projects as achieving career goals, our reasons for following these through, and what happens when we have attained them are far from straightforward. All the practical projects of a finite being have a provisional nature; and, upon attainment, how we reflect on them has a complex character. A clear path to achievement is pleasurable to reflect upon, but the pleasure arises not only from the achievement but from consideration of the alternative—less happy—outcomes that might have arisen, or different and complicating circumstances which might have required a complete transformation of our plans. Of course, if our life-goals are themselves complex, the psychological play between what we have achieved, and what might have happened, and what could happen in the future, becomes all the more powerful.

Of course, there are some people whose ways of existing are extremely static. They like a quiet life organized that which has become habitual to them. However, no human being is able to remember the past, and project the future and counterfactual possibilities with mechanical exactitude. Even the most rigid lifestyle's relation to this horizon has gaps and ambiguities, and all the exaggerations and omissions that go with the imagination's stylizing character. Humans can take a pleasure in rigid social roles, and the formulation of rite and ritual for their own sake. Such things are enjoyed not just as a means of achieving social cohesion, but in their own right—as something one has chosen to do from a range of possibilities.

Crowther argues that, as an expression of freedom, this role-playing is a source of intrinsic pleasure. The following remarks usefully focus his entire position:

> whilst the narrative of the self is inseparably bound up with practical issues, it exceeds these. To comprehend who and what we are is

> not just a case of having specific beliefs and desires, or being able to affirm that 'I think' or imagine. It involves, rather, the understanding and imagination interacting in a freer and more creative way than is required in order to recognize that 'such and such' is the case. Individual ascriptions of self-consciousness are only made possible by virtue of their reciprocal relation with the pleasurable interplay of the developing totality. To be self-conscious entails that we must be able to stand apart from the immediate practical involvements of life. The aesthetic pleasure involved here provides a motive for adopting such a reflective standpoint. It is grounded in practical interests, but insofar as it is not sufficiently analyzable in these terms, it can be characterized as relatively disinterested in character.[16]

Crowther's understanding of disinterestedness will be explored in the next chapter in more detail. However, its use here is to show how Crowther uses it to describe pleasure that is emergent from the relations between phenomena rather than from the practical or physiological gratifications they give rise to. This is why his emphasis on the aesthetic as a basis for the *cohesion of the self* (Crowther's own phrase) does not amount to an aestheticization of it. The aesthetic experience is a pleasure in relations between phenomena. In art, the phenomena are represented rather than real. In experience they are real, but what we enjoy is not just the reality but *also* the way it coheres with other features. To take a pleasure or displeasure in the story of what we have become or are becoming, is a pleasure of reflection that is different from, but nevertheless enhances the broader psychological gratifications of the relations between personal, social, and family existence (and all their physical settings) from which it emerges.

IV The Origins of Symbolic Form

Through the foregoing account, it should have become clear how Cassirer's basic symbolic function is implicit in all the features described by Crowther's approach. The image is a particular that refers to a greater whole—whether that whole is another complete actual or possible individual, or a class of individuals. It is inseparable from our knowledge of objects, and is also at work in all aspects of the self. Indeed, in the narrative plays of self-consciousness, the basic function is embodied in another respect, in that we make sense of our individual moments and situation is respect of their role in a changing narrative or network of narratives, whose nature is itself necessarily connected to the individual character of these moments. Holistic structure is involved.

There is a further consideration from Cassirer that must now be taken account of. In the previous chapter we showed how, for him, symbolic forms are deployments of the basic symbolic function in relation to different objects and relations as presented to us through the basic categories of knowledge—space, time, causality, thing/attribute, number, and magnitude. We will recall that he emphasizes how such forms are progressively articulated through a critical tension arising between individual elements and the character of the whole. It is important to understand the scope of the critical in terms of the relation between Crowther's theory of the narrative structure of the self, and the progressive articulation of symbolic form.

Suppose that one lives in some jungle as a member of a tribe that has no contact with others, and has existed for aeons in a physical environment that provides for basic needs and a little more besides. Such a person thinks of his or her relation to the environment in very much the terms of Cassirer's mythic consciousness. Whilst one's day-to-day routine is very much the same year in year out, one is still part of a world that is full of wonders and spirits amongst which one must find a place. One has hopes and fears in relation to this, and, all in all, even the simple routines of life are given narrative depth by how they play off against the act-association continuum in which they are immersed.

This raises the question of why people ever leave such a state. The main answer, of course, is that challenging physical vicissitudes can create problems that require the invention of new means for dealing with them, perhaps even relocation. To survive one must change and invent. This means that the threat to survival is the major motive for the critical actions that, as we saw above, Cassirer sees as basic to the development of symbolic form.

However, there is more to it than this. The extraordinary diversity of developments within even a single symbolic form suggests that humans take a pleasure in developing such forms for their own sake. Once the world an agent inhabits involves knowledge that ranges beyond the immediate physical environment and the satisfaction of practical needs, then such an agent has a need for narrative meaning. Bare survival, even comfortable survival is no longer enough. Individuals acquire a need for self-development. *The narrative of the self becomes a motive for action as well as a characterization of a major structural aspect of self-consciousness. It is this—over and above the need for survival—that pushes towards the development of symbolic forms.*

Of course, not all individuals are driven in this way, but once the possibility of such development is opened up through the right cultural and physical circumstances, then all that it needs is a few such driven individuals to achieve personal development that progressively articulates the symbolic form or forms with which they are involved. (As we shall see in later chapters, this is of vital significance to art.)

There is a key ethical/ontological pay-off from this. As symbolic forms are developed they deepen the scope of self-understanding in such a way that the critical factor points towards major new possibilities of self-development. An initiation into all the major symbolic forms can offer the agent *critical autonomy*. Through the development of its narratives, such an agent strives towards progressive articulation in a personal sense, by engaging with what symbolic forms offer. And, in doing this, it, at the same time, stimulates the further development of these forms themselves.[17]

Crowther's account of imagination and the self, then, is not only consistent with Cassirer's position, but suggests that progressive articulation begins with the self and its need for narrative development, and that such development is reciprocally enhancing vis-à-vis the progressive articulation of symbolic forms as such.

This raises some final possible objections, and two important transitional considerations. The first concerns the idea of the narrative structure being a source of aesthetic pleasure. If this was so, the objection holds, wouldn't it actually remove the need to create art—we would have aesthetic pleasure, as it were, 'on tap' all the time? And, if this were so, it would justify a second objection. We would be in a constant aesthetic 'high'. Such pleasure would be the norm—which is surely counter-intuitive in relation to the general consensus that, whatever else it is, aesthetic pleasure is something out of the ordinary. A third objection is more radical still. It is the claim that we do not have to introduce the idea of narrative (let alone the aesthetic) as a necessary feature of the unity of consciousness at all. All we need are formal criteria of physical continuity in the same body and the like.

These worries can be dealt with by qualifying Crowther's theories as follows. We have emphasized the importance of the frailty and transience of all our conscious states and the play of imaginings and associations that are central to them. We have emphasized also the involuntary nature of many of these states. The point is that, whatever the pleasure we take in the narrative of our lives, we do not possess the narrative in any enduring form. For this possession to occur, we need to make some artefact or other that embodies the basic symbolic function in an appropriate sensuous form. In this way, our pleasure in comprehending who and what we are becomes more subject to the will. By creating or engaging with an artwork, we receive a more controlled and—in an important sense—free aesthetic pleasure. The aesthetic retains its special character. Indeed, the fact that art or the aesthetic enjoyment of nature answers to something that is so basic to us, enhances their special status.

This connects to the second objection. Crowther can claim that the aesthetic pleasure involved in narrative can be dispositional. Where and when it arises is contingent on the individual and his or her life-story, but for it not to be present, or for it never to be realized in occasional terms, would be counter-intuitive in the extreme.

This also enables the third and more radical objection to be answered. For reasons noted earlier, rational beings have needs over and above merely staying alive, and finding physiological gratification; We would argue that exploring and taking a pleasure in who and what we are is the foundational aspect of these needs. Crowther deals with a similar objection by positing a 'pseudo-experiential' computer that can record audio-visual data, and synthesize how possible states of affairs might appear in the future. All this data is of equal value to the computer. Its 'moments' of 'experience' may be formally different from one another, but there is no hierarchy amongst them. It can replay them exactly as it registered or projected them. Crowther argues that:

> This computer, then, has a formal unity of the sort presupposed by the possibility of self-consciousness, but it is *not* self-conscious. The reason why is that it is too crudely perfect. Its visual 'sights' are discrete, well defined, exactly projectable and thence retrievable. Its moments of seeing form a differentiated continuum but one wherein no elements can outweigh any other. The computer has no reason to be self-conscious. A human being has. . . . It is an embodied *subject* precisely because as well as enjoying the occurrence of such states, it can through language and imagination recall and anticipate them with some degree of pleasure. Concomitant with this, it can also enjoy the means which allow these states of gratification to be effectively realised. These capacities cannot be separated from one another. Neither can they be separated from the key *catalytic* factor, namely that in the struggle for life some Moments have more value than others.[18]

The point is, then, that we cannot make the notion of human experience intelligible without its moments of experience *mattering* to it. The factors that Crowther describes in relation to imagination and the narrative of the self are the basis of this mattering. They are necessary to the 'cohesion of the self'.

Having dealt with the objections, we may now consider the two transitional points mentioned earlier. The first concerns the origins of art as a symbolic form (and relates, accordingly, to the refutation of the first of the objections just considered.) We have emphasized the importance of the act-association continuum as the atmosphere in which all cognitive activity breathes. But images, and our sense of narratives are elusive things—both introspectively, and socially. To fully comprehend ourselves, we need to express image and narrative in a mode that allows them to physically endure in a clarified form, and which makes them available to other persons. The importance of this cannot be stated strongly enough. The lack of relevance of contemporary aesthetics—noted at the beginning of this book—has much to do with its failure to engage with the origins of art vis-à-vis the need to find symbolic expressions for images and narratives, and the embedding of images and narratives in cognitive fundamentals. There is far more to be said about this than we or Crowther have been able to say to date, and this is true also concerning the important question of how art transforms the things it represents. We will, however, be addressing relevant issues as they arise in each of the coming Chapters.

The second transitional point concerns the status of aesthetics itself. It is clear that art and the aesthetic have a fundamental role in our knowledge of objects, self, and other people, and that they can claim the status of being features of a unique form of knowledge. At one time something of this was understood in a restricted way through the connection between Ideal beauty and Platonic ideas. However, this made art and beauty seem transcendentally rather than experientially grounded. It was only from the Enlightenment onwards that this connection was understood in more comprehensive terms as involving cognitive fundamentals concerning the relation between subject and object of experience at different levels.

It is important, then, that in moving forward, we consider how *the rise of aesthetics as a discipline of inquiry is, itself, implicated in the development of aesthetic cognition*. For if art and the aesthetic art are expressions of a unique form of knowledge, then we must have the intellectual resources to explain why this is the case. New and challenging forms of art are necessary, but so is theory—to explain the broader significance of what these forms achieve.

Now Crowther has written much about Kant and other historical figures individually considered, but he has offered no account of the *development* of modern aesthetics itself. Cassirer, however, has addressed the task in some detail. His superb accounts of aesthetics from the Enlightenment to the mid twentieth-century constantly raise ideas and arguments that underline the importance of the aesthetic and the artistic as features that engage with the cognitive fundamentals that we have discussed in this chapter. Indeed, by following where Cassirer leads we will find more areas where his and Crowther's ideas demand to be linked.

Notes

1 In this chapter we draw freely on ideas made in chapters Four and Five of Crowther's *PAP*, and his papers "Imagination, Language, and the Perceptual World: A Post-Analytic Phenomenology", *Continental Philosophy Review* 46, no. 1 (2013): 37–56, and "How Images Create Us: Imagination and the Unity of Self-Consciousness", *Journal of Consciousness Studies* 20, no. 11–12 (2013): 101–123. We do not discuss all the features that operate in Crowther's phenomenology of imagination and his aesthetic of the self—only those most relevant to our study.

2 In this, Crowther is following Kant. See Immanuel Kant, *The Critique of Pure Reason*, trans. N. Kemp-Smith (London: Macmillan, 1973), 111.

3 Gareth Evans, *Varieties of Reference* (Oxford: Clarendon Press, 1982), 176.

4 Edward S. Casey, *Imagining: A Phenomenological Study*, 2nd Edition (Bloomington: Indiana University Press, 2000); J.-P. Sartre, *The Imaginary: A Phenomenological Psychology of the Imagination*, trans. Jonathan Webber (London: Routledge, 2004); Mark Johnson, *The Body in the Mind: The Bodily Basis of Meaning and Imagination* (London: University of Chicago Press, 1990) (especially Chapter 6). Each of these thinkers picks out the features that Crowther

describes, but none of them assign special emphasis to the quasi-sensory aspect, or link imagination to the unity of the self.

5 Colin McGinn, *Mindsight* (Harvard: Harvard University Press, 2004).
6 Crowther, *PAP*, p. 76.
7 Cassirer, *EM*, p. 51.
8 Crowther, "How Images Create Us . . . ", p. 109.
9 This emphasis on instability is the main feature that is common to all the main recent philosophical accounts of the imagination noted earlier.
10 See Alasdair MacIntyre, *After Virtue: A Study in Moral Theory* (London: Duckworth, 1985), Chapter 15, 204–225. An interesting cognitive psychological approach to the relation between narrative and the momentary aspects of experience can be found in Chapter 4 of David Pillemer's *Momentous Events, Vivid Memories* (Cambridge, MA: Harvard University Press, 1998).
11 Crowther, "How Images Create Us . . . ", p. 11.
12 Crowther, *PAP*, p. 88.
13 Ibid., pp. 88–89.
14 Ibid., pp. 95–96.
15 Ibid., pp. 94–95.
16 Ibid., p. 97.
17 The concept of *critical autonomy* is the main theme of the Conclusion to Crowther, *PAP*.
18 Crowther, *PAP*, p. 8.

3 Cassirer on the Rise of Aesthetics

The history of aesthetics since the Renaissance has recognized that beauty and art involve higher values—they do not just satisfy, they can be cultivated, and through this cultivation, humankind engages with truths about its relation to the cosmos, and the need for finite beings to find a place in relation to it. Through beauty and art, we can contemplate and aspire towards that which is beyond the transitoriness of the finite world.

In the Enlightenment, however, this ethos of transcendence is reinterpreted—mainly through responses to societal developments, the diversification and advance of knowledge and technologies, and, of course, through a recognition of new complexities in the arts. Cassirer's historical analyses offer a way of understanding the unities that underlie this reinterpretation. In particular, his account foregrounds their first presentations of the aesthetic's connection to cognitive fundamentals concerning the reciprocity of the human subject's understanding of the objective world, and itself. It is this *aesthetic cognition*, as we have termed it, that requires the development of aesthetics as a discipline. For when symbolic forms articulate reality in increasingly complex ways, philosophy must simultaneously focus and extend its analytic techniques in order to comprehend this complexity.

In this chapter, therefore, we will analyse Cassirer's arguments in *The Philosophy of the Enlightenment* and *Kant's Life and Thought*. In the course of this, we will see how aesthetic cognition emerges and, in the manner and content of its emerging, creates aesthetics as a discipline.[1]

I The Development of Enlightenment Aesthetics

Cassirer suggests that there had always been a close link between philosophy and literary criticism. In the Enlightenment these links gradually become formalized in the notion of a systematic philosophical aesthetics. This involves not just an analysis of art but is an endeavour, which seeks to correlate the content of philosophy and art as forms of knowledge. But art is especially difficult to analyse. The problem here emerges as an opposition—not only between art and philosophy but between in the broadest terms the objective and subjective components of art itself. As Cassirer puts it in relation to both these factors:

> Whether it is the dispute between reason and the imagination, the conflict between genius and the rules, the foundation of the sense of beauty in feeling or a certain form of knowledge: in all these syntheses the same fundamental problem recurs. It is as if logic and aesthetics, as if pure knowledge and artistic intuition, had to be tested in terms of one another before either of them could find its own inner standard itself in the light of its own relational complex.[2]

This complex process finds its first great resolution (although not a definitive one) in the work of Kant. But before Kant's clarification could be achieved, various logical phases in the progressive articulation of the fundamental opposition had to be worked through. Cassirer traces these logical moments in terms of four basic phases; a) classical aesthetics and the objectivity of the beautiful b) taste and the trend towards subjectivism c) intuitive aesthetics (i.e., Shaftesbury and his legacy, and d) the systemisation of aesthetics in German thought (most notably Baumgarten). I shall now endeavour to explicate these phases in turn.

The classical tradition in Enlightenment aesthetics finds its first major exemplar in Batteux's *The Fine Arts Reduced to a Single Principle*. For Cassirer the ambition expressed in this title is part of the Cartesian legacy of the yearning for a universal knowledge reducible to clear and distinct first principles. There will be first principles of art as there are for nature, based on some general principle of imitation. But there is a special problem for the artistic version of such a principle. In scientific understanding, imagination must be suppressed, but to do this in relation to art might at first sight seem to negate the whole basis of artistic meaning—the more so insofar as, traditionally, the execution of a perfect work requires accomplished technique, and an innate talent or power of imagination.

Boileau deals with this problem by considering that whilst fancy and imagination are integral to conception and the process of creation, these must be totally subordinated in the finished artwork to an objective law of beauty. The artist does not invent this, but discovers it through rational investigation into the nature of things. In more general terms, the artist must articulate and comprehend laws governing the internal structure and relation of the different genres of art. In this way the genius is protected against arbitrariness and is guided in his selection of forms. Given an appropriate form, the artist can then exhibit powers of expression in its execution. This will constitute originality (but not of course originality at any price).

For Cassirer, such Classicism offers a first principle of objective beauty, but it is in methodological terms flawed. In classical theory:

> [t]hey all strive for simplicity, propriety, and "naturalness" of expression, but they derive the standard of the natural without doubts of

> or scruples from the world in which they live; they base this standard on their immediate environment, on habit and tradition. At this point the power of abstraction, with which the founders of classical doctrine are endowed, begins to fail; critical reflection is supplanted by naïve credulity, by a veneration for all those things which were empirically given in the intellectual and artistic culture of the seventeenth-century.[3]

Classicist perception of form is actually based on a socially determined notion of propriety. Eighteenth-century thought disrupts this view because it comes to hold that different varieties of art are considered appropriate for different ages. As Cassirer declares '[a] theory is now required which suits the variety and changeability of aesthetic phenomenon. The trend is from a mere formula to a knowledge of the real form of the artistic.'[4]

Cassirer negotiates this new demand in terms of 'Taste and the Trend Towards Subjectivism'. Again, the tendency, which Cassirer is describing, is connected with bigger shifts of symbolic form. In this case it is an interest in the psychology of human nature. As Cassirer says,

> The psychological approach, which finds its origin and basis of the beautiful in human nature, by no means intends to give way to an unlimited relativism, or to recognize the individual subject as the absolute judge concerning works of art. This approach considers taste as a sort of "sense" shared by all, and it begins its formulation of the problem with the question of the nature and possibility of such a "common sense" (*sensus communis*).[5]

Glimmerings of such a view are already apparent in some classicists such as Bouhours (writing in the 1680s). He opposes Boileau's notion of 'correctness' with that of "*délicatesse*". This emphasizes lightness and flexibility of thought, unexpectedness, shades and nuances of meaning. Such a notion amounts as Cassirer puts it to 'as it were, a new organ'.[6] What is an emphasis in Bouhours is made into a central concern by Dubos. In his work it is not a case of finding a place for imagination and feeling, but of asserting their fundamentality. He establishes introspection as the defining principle of aesthetics. In Cassirer's words,

> [t]he nature of the aesthetic cannot be known by mere concepts, and the theorist in this field has no other means than communicating this insight to others and of convincing them of its truth than to appeal to their own inner experience. The immediate impression, with which all formulation of the aesthetic concepts must start and to which it must constantly refer, cannot be replaced by any deductions.[7]

Taste—the aesthetic response—has an immediacy that parallels pure acts of perception such as seeing and hearing, tasting and smelling.

Hume takes this position through to a logical conclusion. What gives his position special power is his reinterpretation of reason itself as the "slave of the passions". In the context of aesthetics, this totally affirms the primacy of imagination and subjective taste. The celebrated essay 'Of the Standard of Taste' makes this clear. For Hume, value judgments are concerned not with things and their nature *per se* but rather in the *relation* between object and human powers of perception, feeling and judgment. The judgment of taste is "true" in every instance, without having to be the "same" in every case. Such considerations give the aesthetic judgment an advantage over logical ones:

> For in resisting all false generalisations and in making assertion not about objects as such, but about our relations to them, the aesthetic judgment is enabled to achieve that conformity for which the sciences of the objective world strive in vain. The individual subject can of course never presume to be a judge of things, but he is the only possible and the only qualified judge of his own states of being. Yet the aesthetic judgment is intended precisely as an expression of these states. Hence it can accomplish more because it attempts less.[8]

We, therefore, are able in a sense to make objective judgments about beauty insofar as beauty is not an object but a condition within ourselves. Now this might seem to eliminate the possibility of universal validity from the aesthetic judgment. However, for Hume it does not rule out a standard of empirical uniformity. Indeed, whilst philosophical systems rapidly supersede one another in the passage of time, great works of art maintain their audience much more successfully. As Cassirer notes '[i]t would be more correct to say that they bridge the centuries and become the best evidence that if human thinking changes, then at least human feeling and capacity for aesthetic impressions remain basically the same'.[9]

Hume's position on this does not satisfy other eighteenth century thinkers of an empiricist inclination. Diderot, for example, holds that taste is subjective in that it is grounded in feeling, but that it has nevertheless, an objective dimension through individual judgments being informed by innumerable past experiences of beauty. As Cassirer says '[o]ur feeling for the beautiful results from the accumulation of all these observations and experiences in our memory and from their fusion into a new general impression'.[10] However, this leaves open the question of both why we find things beautiful, and how those responses are to be separated from our experience of beautiful forms, which have been encountered in practical contexts. Again in Cassirer's words,

> Wherever Diderot seeks not only to describe but also to explain the beautiful, he finds this possible only if he considers the beautiful dependent on the "true" and merely as a disguised form of the latter. Only now the norm of truth has changed; its content is no longer based on a

> priori propositions, on general and necessary principles, but on practical; experience, on daily routine and utility. Both types of explanation fail, however to account for the peculiar meaning and value of the beautiful; for the standard employed in both cases is on a different plane from that occupied by the pure phenomenon of beauty.[11]

The route to the "pure phenomenon of beauty" is to be found in another direction—that of what Cassirer calls Shaftesbury's "intuitional aesthetics". The significance of Shaftesbury is that, whilst having been a pupil of Locke, Shaftesbury's philosophical endeavours are shaped by sources, which enable him to by-pass the mainstream of eighteenth-century British empiricism. These sources are such classical thinkers such as Plato, Aristotle, Plotinus, Seneca, Marcus Aurelius, and Epictetus. He also has a much broader approach to the problem of beauty as such—one which ranges far beyond the specific connection with art. In general terms, he gives the classicist dictum "beauty in truth" a complex meaning. Cassirer notes that:

> To him [i.e., Shaftesbury] "truth" signifies rather the inner intellectual structure of the universe, which cannot be known in terms of concepts alone or grasped inductively by means of an accumulation of individual experiences, but which can only be immediately experienced and intuitively understood. This form of experience and of intuitive understanding is available in the phenomenon of the beautiful.[12]

For Shaftesbury, the formative principle of beauty also allows the general formative principle of nature and the spiritual world to be disclosed to contemplation. The immanent order, which sustains the entire universe, finds a focussed means of expression through the beautiful. The beautiful is in essence, the contemplation of creative cosmic order.

This notion is linked by Shaftesbury to art through the idea of *genius*. Artistic imitation does not address in the surface of things and mere appearance, but occupies a different dimension. In Cassirer's paraphrase 'it imitates not merely the product by the act of producing, not that which has become, but the process of becoming. The ability to immerse itself in this process and to contemplate it from this standpoint is, according to Shaftesbury, the real nature and mystery of genius'.[13]

The linking of genius to the comprehension of formative creative power is for Cassirer a great achievement in the history of aesthetics. It is a moment whose full potential is realized (as we shall later show) in Kant's aesthetic theory. The line of transmission from Shaftesbury is complex and proceeds chiefly via Francis Hutcheson. His notion of beauty is constructed on the analogy with sense perception. There is an inward sense of beauty, which is independent of, and different from the external senses. A person, who is not gifted with this inner susceptibility to harmony and order, cannot acquire it through processes of learning.

In the terms of Cassirer, however historically significant Hutcheson may be his theory amounts to a "blunting process" performed upon Shaftesbury's position.[14] Genius is, in effect, equated with the gift of receptivity and good taste. Worse still, by construing beauty as recognizable exclusively through an independent sense, he divorces it from that more basic formative power which, for Shaftesbury is not only the essence of beauty but also the natural and spiritual order. (The full significance of why this is such a problem for Cassirer will be considered later in our discussion of his treatment of Kant's aesthetics).

A superficial parallel to Shaftesbury's dynamic concept of beauty is also discerned by Cassirer in the work of Dubos:

> They are opposed to all attempts to confine the beautiful to strict immutable rules; they admit that genius has the right and power to break the tablets of the law and to establish new rules by virtue of its own authority. They also turn against all attempts to comprehend the nature of the beautiful by mere reasoning, by means of discursive definitions and analytical distinctions. They advocate an immediate knowledge of the beautiful.[15]

Where Shaftesbury and Dubos differ is in their positive standpoints 'for the former lies in the process of pure creation, whereas for the latter it is sought in the region of certain modes of reception and conception which admit of no further derivation'.[16] Dubos identifies any dynamic feature of the aesthetic experience purely with the receptive dimension of feeling. According to Cassirer this threatens to reduce the work of art to a mere spectacle where, so long as it entertains and intensifies the audience's feelings, it makes no difference what the means are by which the effect is achieved. This stands in complete contrast to what Cassirer describes as Shaftesbury's '. . . most important individual contribution to aesthetics'—namely the linking of beauty to *disinterested pleasure.*[17] For Shaftesbury whilst all beauty is truth, this does not signify the dependence of the former on the latter. They are as Cassirer has already showed, logically independent aspects of a deeper formative power. It is this power which, for Shaftesbury is the basis of art. In Cassirer's words:

> The genuine artist does not laboriously collect the elements of his creation from nature; he copies rather a model which stands before him as an original and indivisible whole. All this imaginative model is no mere illusion; on the contrary, it is sure to agree with the general truth of things, if not with their factual reality.[18]

The artistic genius, which accomplishes such achievement, comprises a natural gift for formative activity. For Cassirer, this suggests that Shaftesbury rightly regards the ego as 'a primary whole and an indivisible unity' however diverse its contents may be.[19]

Given a conception of art such as this, disinterested contemplation is *prima facie* a much more appropriate response than the entertainment and intensification of feeling posited by Dubos. However, Cassirer's treatment of the link between aesthetic experience and disinterested pleasure is somewhat cursory, and we will need to return to it a little further on.

Having established the momentous contribution of Shaftesbury, Cassirer further suggests that the aesthetic theory of Edmund Burke also has a decisive role to play in the development of Enlightenment aesthetics. The reason for this is not only the distinction between the beautiful and the sublime, which Burke makes in his theory, but also the way that he makes it. As Cassirer paraphrases him '[t]he beautiful unites, the sublime isolates; the one civilizes by teaching the proper forms of social intercourse and by refining morals, the other penetrates to the depths of our being and reveals these depths to us for the first time'.[20] In Burke's theory in other words, the sublime is not simply identified with Longinian qualities of excellence and lofty vision, but with an aesthetic experience which 'removes the boundaries of the finite'.[21] In this experience, the ego does not experience the removal of these boundaries as a destruction, but rather as a liberation and exaltation. For Cassirer, the Burkean ideas are of the greatest historical significance. As he puts it,

> With more rigor that heretofore this theory distinguishes essence from appearance, nature from custom, the substance and true depth of the ego from its merely accidental properties and relations. The problem of genius and the problem of sublime are now moving in the same direction, and they become the twin bases upon which a new and deeper conception of individuality gradually evolves.[22]

So far, then, Cassirer has identified a classicist-objectifying trend in Enlightenment aesthetics, and a more empiricist reception-oriented reaction to it. In Shaftesbury and Burke, he further identifies a pair of approaches which break down the barriers between objectivist and empiricist approaches by linking aesthetic experiences to a more active and formative conception of the self, which (in Shaftesbury's case at least) is the manifestation of a more universal dynamic.

In the final sections of his analysis of Enlightenment aesthetics, Cassirer investigates the way in which these (and other themes) intertwine and are eventually systemized in eighteenth-century German aesthetic theory. His starting point for this is a section entitled 'Reason and the Imagination: Gottsched and the Swiss Critics'. Here Cassirer centers on the fact that [German] '[a]estheticians did not attempt to free the imagination from the predominance of logic; they wanted to discover a logic of the imagination'.[23] This strategy finds its best expression in the antagonism between Gottsched and the Swiss critics Bodmer and Breitinger. All the parties involved in this dispute were sympathetic to the systematic rationalist philosophies of Leibniz and his follower Wolff. They put this rationalist foundation, however, to very different uses. In Cassirer's words for Gottsched '[t]he "theorem", the theoretical or moral truth, thus comes first; the poetic actions follows simply to exemplify this truth in a concrete case'.[24] The Swiss critics reverse this emphasis:

> To be sure, they have by no means renounced a moral intention; they stress such an intention repeatedly. But this poetic moral is now to be realized not by way of reason but by way of imagination. The task of poetry—is to stir the emotions, but the pathetic is not of course their only and highest goal. Stimulation of the imagination is rather to prepare the way for rational insight and to predispose the listener to such insight. What the mere concept and abstract doctrine cannot achieve is to be accomplished by the proper choice of metaphor and poetic imagery.[25]

One signal advantage of the Swiss critics' position is that they seek to derive a system of rules and norms for poetry by analysing existing works of a high order, rather than prescribing them in advance as a system that great works must fit. Again Cassirer draws an analogy here between aesthetics and contemporary science '[a]s the physicist finds mathematical precision among sense data, so the art critic seeks in the creations of the imagination that which is necessary and above all caprice. He begins with direct contemplation and remains faithful to this method; but in it he discovers the specific form of determinism and that "demonstrative certainty" of which such contemplation is capable'.[26] The Gottsched/Swiss critics' antagonism centers, in effect, on how aesthetic criticism is to be effectively systemized. For Cassirer, it is Baumgarten who brings about the most effective systemisation of this sort before Kant. He, in effect, lays the philosophical foundations of aesthetics as a discipline. Cassirer sets out the basics of this in an important passage as follows:

> Aesthetics would not be a science and could never become one if it confined its activity to giving technical rules for the production of the works of art or to making psychological observations concerning its effect on the spectator. Such activities belong to that kind of empiricism which is the exact opposite of real philosophical insight, and from the view point of method, they form the sharpest conceivable contrast to such insight. The philosophical content and significance of every science is derived from the understanding of its meaning within the whole body of human knowledge, and from the position it occupies in this whole. Every science must fit into the general genus of knowledge; but within this genus it must fulfil this task in a manner characteristic of the given science.[27]

In the first part of this passage Cassirer is of course identifying the weakness, which he had identified in classicist approaches to the objectivity of beauty and empiricist reception based approaches. The remainder of the passages indicates why he has been so favourable to Shaftesbury. For Shaftesbury shows both the logical distinctiveness of beauty and its relation to the broader scheme of things. In effect, he conceives beauty in terms of the basic symbolic function. It is by systemizing this insight with much greater clarity that Baumgarten is able to make his important achievement.

The steps to this are traced by Cassirer as follows. For Baumgarten, the sphere of reason encompasses order and concepts in general, no matter what their medium of embodiment. Following Leibniz he also holds that the search for clear and distinct knowledge involves the analysis of complex phenomena into the ultimate element from which they are constituted. From this analysis it is possible to reconstruct the phenomena in question as logically clarified objects of knowledge. Now within this framework beauty has a special significance. It is a mode of "sensitive knowledge"—the embodiment of "phenomenal perfection". Cassirer notes 'such perfection, however in no sense corresponds to that perfection for which logic and mathematics strive in the development of their "distinct concepts". It asserts itself side by side with the perfection of logic and remains an irreducible and independent element'.[28]

The ground of this independence is the fact that the procedure of logical analysis from whole to constituent elements does not offer a sufficient basis for the logical clarification of beauty. To illustrate this Cassirer notes that:

> If, in accordance with the method of exact science, we explain the phenomenon of color by resolving it in a pure process of motion, then we have not only destroyed its sense impression but we have also robbed it of its aesthetic significance. The reduction of colour to its physical concepts annihilates, as it were with one blow, its whole significance as an artistic means of expression, its whole function in the art of painting.[29]

The fact that aesthetic phenomenon is not amenable to analytic reduction has its own positive significance. This is illustrated by means of the contrast between poetic and logical discourse. Baumgarten assigns 'light', 'clarity', 'abundance', 'magnitude', and 'truth' to the former, whereas the formation of logical concepts is geared towards generalization and abstraction, and is brought at the expense of sensible determinacy. Generality and determinacy lie in opposite directions, Baumgarten's notion of beauty as sensible knowledge, however, is able to connect them. In Cassirer's words,

> Aesthetics bridges this chasm; for its "truth" cannot be found beyond or in opposition to concrete qualities, but it can be realized only amid and by virtue of such qualities. Beauty requires not only intensive clarity, as do scientific concepts; it possesses also extensive clarity. Intensive clarity is achieved when we succeed in condensing the manifold of perception to a few fundamental definitions in which the essential nature of this manifold is recognizable. Aesthetics' extensive clarity, however, does not tolerate any such conceptual reduction and concentration. For

> the artist seeks to portray the full scope of intuitively perceived reality; he seeks to encompass both centre and periphery in a single glance.[30]

> Artistic genius, indeed, looks not to the cause of the phenomena but rather '. . . endeavours to comprehend phenomena in their totality and in their purely immanent mode of existence and to fuse them into one lucid picture.[31]

It follows, then, that for Baumgarten, art and more general logical or scientific knowledge share a linguistic character, but employ it in different direction. The artist endows the "cold symbols" of linguistic or representation forms with the "life of knowledge" (*vita cognitionis*). Hence whilst Baumgarten regards the science of the beautiful as a logic of "lower cognitive forces", beauty and art are regarded as an indispensable dimension of being human. Cassirer indeed, sees this as comprehending the more general relation between philosophy and art in the proper way. In Cassirer's words '[t]he philosopher is and remains akin to the artist in a fundamental feature of his thinking, in his striving for totality; and, if he cannot compete with the artist in the creation of the beautiful, yet he can seek a knowledge of the beautiful and by this knowledge, by means of systematic aesthetics, he can bring to completion his own world picture'.[32]

In addition to this, Cassirer sees Baumgarten in terms of one other decisive achievement. In assigning a necessary place for beauty as *sensible* knowledge in this overall metaphysical scheme, Baumgarten allows a factor essentially tied to mankind's *finitude* to assert its own distinctive character within that scheme. In effect, Baumgarten pushes our conception of man in the direction of a philosophical anthropology. As Cassirer observes:

> Thus, in the severe discipline of German school philosophy the same thought emerges which we have already found operative in the philosophy of morals and religion, and the law of politics, in the age of the Enlightenment. The Enlightenment gradually learned to do without the "absolute" in the strictly metaphysical sense, without the idea of "God-like knowledge". Instead appears a purely human ideal which the age seeks to define and to realize with increasing precision and rigor.[33]

This completes our exposition of Cassirer's analysis of Enlightenment aesthetics. Before looking at Kant, it is worth considering some interpretative issues, which are evoked by Cassirer's discussion.

II Critical Issues Arising from the Rise of Aesthetics

As the foregoing account shows, Cassirer covers over a hundred years of thought in European aesthetics. In view of this scope it is not surprising that there are some omissions in his account, and that his interpretation of some thinkers and concepts require more elaboration than he offers. We now consider a few such problematic issues. In the first place, it is surprising that Cassirer offers no analysis of Addison's 'Essays on the Pleasures of the Imagination' (first published in 1712) although these are generally accepted as the first formulation of aesthetic experience, which broadly breaks free of the Classicist orthodoxy. The omission is all the more surprising because they were also translated into German in 1745, and exerted a great influence in that context.

Rather more serious is Cassirer's misunderstanding of one of the British thinkers he does address, namely Edmund Burke. The significance of Burke for Cassirer is that his way of theorizing the sublime is a liberation of the self. In discussing Burke, Cassirer asserts that '[t]he sublime consists precisely in this transcendence and derives its effectiveness from it'.[34] Indeed '[t]hat we do not succumb to the terrible, but that we maintain ourselves against it and that we actually feel an exaltation and intensification of our powers in its presence—these are the elements of the phenomenon of the sublime and the basis of its deepest aesthetic effect'.[35]

Unfortunately, this is *not* Burke's position. Burke was strongly influenced by Dubos, and his account is, accordingly, fundamentally a reception-oriented one. The sublime is a powerful emotion insofar as it arises from the experience of mild pain or terror from a position of safety. But this does not involve any kind of affirmation of the self or transcendence. In Burke's words,

> if the pain or terror are so modified as not to be actually noxious; if the pain is not carried to violence, and the terror is not conversant about the present destruction of the person, as these emotions clear the parts of a troublesome encumbrance, they are capable of producing delight; not pleasure, but a sort of delightful horror; a sort of tranquillity tinged with terror.[36]

Burke explains the sublime passion, in other words, in fundamentally *causal* terms. The odd mild pain or shock can- under the circumstances described by Burke—stimulate those finer parts of the body to which prolonged inactivity is injurious.[37]

Even in his misreading of Burke, however, Cassirer is prophetic. It is clear (as we shall show later) that what Cassirer is looking for in Burke's theory of the sublime is the kind of idea, which Kant proposes in his theory. This allows us to say that whilst Cassirer's reading of Burke is in error, this error is not of structural significance. What he wants from Burke in relation to the sublime's function in art and the aesthetic (understood as symbolic form) will later be provided by Kant.

There is another problem however, which might appear to raise a genuine structural difficulty for Cassirer. Earlier we noted how he describes Shaftesbury's link between beauty and disinterested pleasure as that thinker's 'most important individual contribution to aesthetics'. But it is curious how little Cassirer comments on the relation of art, the aesthetic and disinterestedness. Indeed, if the notion of disinterestedness is of so much significance why does it not make any further appearance in Cassirer's account? Many writers (notably, Osborne, Beardsley and Crowther) have seen disinterestedness as a key characteristic of all aesthetic judgments, and as the major factor in the development of eighteenth-century aesthetic theory. How, then, can Cassirer be justified in his cursory treatment of disinterestedness?

A shallow response might be to say that we should wait for the truth about disinterestedness—which will happen when Cassirer deals with the thinker who gives it most 'clout', namely Kant. This response, however, would be against the grain of Cassirer's general approach. As we have shown, for him the elements in a symbolic form—even if they are definitive of it—do not simply happen from nothing, they are progressively articulated. Hence we are entitled to demand more of Cassirer on the developing significance of disinterestedness in eighteenth-century aesthetics.

There is however another possible response. The key to it can be found in the way that Crowther characterizes the problem of disinterestedness in the following passage.

> Most commentators have taken this to be a purely psychological question, and have regarded disinterestedness as an attitude taken up to facilitate the cognition of aesthetic qualities, or as a phenomenological characteristic of a mode of feeling consequent upon such cognition. The notion of disinterestedness has, however, a more crucial and logical application, in that, if our appreciation of some phenomenon is found solely on grounds which do not necessarily presuppose reference to its (in the most general sense) practical significance, then, logically speaking, we would be entitled to characterize our experience of such a phenomenon as disinterested.[38]

The key significance of Crowther's point for our purposes is that it inadvertently picks out the fundamental basis of Cassirer's goal in approaching eighteenth-century aesthetics. Of course Cassirer gives due importance to the psychological effects and character of art and the aesthetic, but his fundamental aim is to emphasize how eighteenth-century thought (most notably in Shaftesbury and Baumgarten) gradually clarifies the *grounds* of such effects and character—the factors which make disinterestedness possible. These grounds are to be found essentially in the artwork as a realization and completion of a formative principle.

This enables us to say that we should see Cassirer's relegation of disinterestedness as the hallmark of an *original* reading of the Enlightenment. Rather than viewing the development of aesthetics through the convenient 'disinterested attitude' spectacles of twentieth-century formalism he is showing us that there is an alternative perspective on this development—one which gives a logical priority to the artwork and its creation, rather than to the conditions and nature of its reception. In a sense this is the reason for which Cassirer's misunderstandings and omissions can be compensated. Whatever his errors, he formulates four logical strands—objectivity, subjective response, formative principle, and systematization—whose interweavings allow the emergence of a mature and highly workable philosophical aesthetics in Kant.

It may, of course, be puzzling that Cassirer ends his analysis of the Enlightenment aesthetics with Baumgarten, and offers only cursory references to Kant. The reason for this is that Kant's appropriation of the Enlightenment legacy, gives it a new significance and marks, therefore, a movement beyond. We now, accordingly, consider Cassirer's approach to Kant's aesthetics.

III Kant's Aesthetics

Cassirer's major investigation of Kant's aesthetics is to be found in his book *Kant's Life and Thought* of 1918.[39] One of the extraordinary features of this book is the central significance, which it gives to the *Critique of Judgment* in Kant's critical philosophy. The reason why Cassirer assigns so much importance to the third *Critique* is determined by a particular (and highly original reading) of Kant's relation to the history of philosophy. It is of course, customary to see the Critical philosophy as fundamentally a response to Hume's scepticism concerning the necessity of the causal relation, and, correlated with this, Kant's desire to give philosophical ratification to the scientific achievements of Newtonian mechanics. Cassirer, however, identifies a further agenda on Kant's part.

Its origins lie in the historical legacy of neo-Platonism. In Plotinus's system, the development of the world as a whole is to be explained as a primordial 'emanation' whereby the intelligible ground of the universe gives form to the sensible world through determinate phrases and stages. The structure of the sensible world is, accordingly, only intelligible through us being able to articulate it as an effect of the fundamental formative cosmic principle. This is disclosive of two avenues of knowledge, one corresponding to the workings of the finite intellect, and the other reflecting the standpoint of the unconditional cosmic intellect. The former is dependent on the latter. In Cassirer's words,

> Everywhere that we seem merely to array particular with particular, to pass from the special case to the genus and to divide this once more into species, a prior, implicit, deeper assumption holds sway. Without the assumption of the world as a totality possesses a pervasive, all embracing logical structure, so that one can find no element in it which is totally unconnected with all else, sheer empirical classification and comparison would lose all force.[40]

Cassirer argues that this metaphysical assumption of all-embracing *logos*, even informs the rise of seventeenth and eighteenth century science and philosophy, insofar as they privilege mathematical understanding. Cassirer observes that:

> Wherever we have before us a whole, not of things but of knowledge and truths, the same question is posed. Every such logical whole is at the same time a logical construction in which each member conditions the community of all the rest, just as it is simultaneously conditioned by them. The elements are not arranged adjacent to one another but exist only because of one another; within the complex the relation in which they stand necessarily and essentially belongs to their own logical existence.[41]

The 'logos' which sustains the elaboration of mathematical truth is of supreme importance for Kant's Critical epistemology and beyond. The task of *The Critique of Pure Reason* is to show how it is possible for the realm of appearance and perception to be organized in the unity of an objective world. Through his transcendental arguments he shows that experience must have spatiotemporal and causal stability and intelligibility, through the application of space and time as forms of intuition and the pure 'categories' of the understanding. Now when the notion of the *a priori* foundations of teleological judgment is introduced in the *Critique of Judgment* Kant presents it as a unifying principle that bridges the gap between nature and freedom. Freedom is expressed through acts of rational agency, which adopt rules of activity so as to achieve specific ends. This harmonizes with our theoretical knowledge of nature insofar as in order to give systematic unity to our knowledge of nature (over and above the unity provided by the categorical order). We act *as if* the order of nature is one, which is designed so as to facilitate the aims of theoretical knowledge. The relationships between part and whole, and genera and species in organic life, is taken to be purposive. Individual forms are 'final' in relation to more general structures or 'ends.'

Kant, then, is proposing that our consideration of the complexity and diversity of natural forms, we are compelled to interpret them on the analogy with artifice in order to render them intelligible. The teleological judgment is not constitutive of experience—we could have experience of an objective world without it. But it is necessary to the degree that we could not have *systematic theoretical* knowledge without it. It is *a priori* in a *regulative* sense.

We are now in a position to show what is so original about Cassirer's reading of this doctrine of Kant's. Kant himself discusses the teleological judgment primarily in reference to the organic forms of nature. At the same time, he also allows it a rather broader function. This is shown in the following remarks.

> Natural teleology forms part of physics, and is a method applicable to the solution of the problem of physics. Arithmetical and geometrical

> analogies, also universal mechanical laws, however strange and worthy of our admiration the union in a single principle of a variety of rules apparently quite disconnected may seem, have no claim on that account to rank as teleological grounds of explanation in physics. They may deserve to be brought under review in the universal theory of the finality of the things of nature in general, but, if so, this is a theory that would have to be assigned to another science, namely metaphysics.[42]

It is important to note wariness on Kant's part. He allows teleology a role in those laws connected with space-time and causality, but this teleology is of a 'technical' sort—rather than the intrinsic variety in terms of which we read organic life. To do justice to technical teleology would require situating it in a broader metaphysical theory. It is precisely this contextualization which Cassirer's interpretation provides. In the foregoing quotations concerning logical order and mathematics, Cassirer is able to identify the precise way in which teleological structure informs the formal comprehension of spatial knowledge, and, by extension, temporal and causal knowledge too. He takes Kant at his word. Space, time and the categories are not simply innate ideas but *active principles* for the construction of objective experience. They are rules geared towards *the unity of experience as an end*. As Cassirer says in relation to the comprehension of geometric space,

> this yields precisely that cohesiveness and correlation of parts which constitutes the essential factor in Kant's concept of purposiveness. Purposiveness thus can be found not only in accidental formations of nature but also in the strictly necessary formation of pure intuition and pure concept.[43]

What Cassirer's reading suggests is that the distinction between principles which are necessary conditions of having experience (i.e., constitutive) and those which are concerned with the ordering of knowledge (i.e., regulative ones) both have an underlying dynamic unity of teleological unfolding. There is an implicit *logos*. What Kant's critical standpoint does is to totally rethink the legacy of neo-Platonism by making *logos* into an epistemological and methodological principle rather than an ontological truth about the nature of the universe. In Cassirer's words,

> Even the thoroughgoing unity of the particular forms of actuality and the particular laws of experience may thus be regarded as if an understanding (though not our own) had produced them for the benefit of our faculties of cognition, so as to render possible a system of experience under particular laws of nature. But we do not thereby assert that this forces us really to postulate each an understanding; rather, the judgment in this case legislates only for itself, not for nature, since it is mapping a course for its own reflections.[44]

This reading of Kant as a radical reformer of the neo-Platonist tradition also emerges with full force as part of Cassirer's reading of Kant's aesthetic theory. Cassirer suggests that:

> The speculative aesthetics that grew out of the circle of the Florentine academy, thence to have its effects from Michelangelo and Giordano Bruno down to Shaftesbury and Winckelmann, is an extension and development of the fundamental motif sounded by Plotinus and Neoplatonism. In this view, the work of art is only one particularly notable specimen of that "inner Form" on the cohesiveness of the universe as a whole rests. Its composition and articulation are the immediately intuitive isolated expressions of what the world *as a whole* is.[45]

The essence of Cassirer's reading of Kant's aesthetic theory is to transpose this neo-Platonic conception from an ontological to an epistemological level where

> [t]he work of art is something singular and apart, which is its own basis and has its goal purely within itself, and yet at the same time in it we are presented with a new whole, and a new image of reality and of the mental cosmos itself. Here the individual does not point to an abstract universal that stands behind it, but it is this universal itself, because it comprises its substance symbolically in itself.[46]

Cassirer's route to this general claim is via Kant's general theory of the beautiful and the sublime. Like Kant himself, Cassirer commences with an analysis of the disinterestedness of the pure aesthetic judgment. In making such a judgment:

> The content here appears in that qualitative perfection which requires no external completion, no ground or goal lying outside itself, and it brooks no such addition. The aesthetic consciousness possesses in itself that form of concrete realization through which, wholly abandon to its temporary passivity, it grasps in this very fleeting passivity a factor of purely timeless meaning.[47]

Hence the distinctive characteristic of the aesthetic judgment is an orientation towards formal structure to the exclusion of all other factors, which would link it to practical interests and theoretical knowledge. It is this orientation towards the structure of appearance alone that gives such judgments their logically distinctive 'disinterested' character. An aesthetic form has the appearance of purposiveness in the relation of its phenomenal whole and parts, but this structure is found valuable in itself, rather than for any more general end which it might exemplify. Cassirer summarizes all these points as follows.

> For purposiveness means, as has been shown, nothing other than individual creation, which displays a unified form in itself and in its

> structure, while purpose means the external determination which is allotted to it. A purposive creation has its centre of gravity in itself; one that is goal oriented has its centre external to itself; the worth of the one resides in its being, that of the other in its results. The sole function of "disinterested pleasure" is to bring this state of affairs, considered subjectively, into thought.[48]

This raises the question of why the mere appearance of purposive structure should be so valuable. The answer is that the perception of such 'formal finality' brings the faculties of imagination and understanding in a general and mutually stimulating interaction. Any empirical judgment involves the co-operation of imagination and understanding in a fixed relation bound by a definite concept. Finality of form, however, engages them in a less rigid way, which enlivens their general capacity for interaction. In Cassirer's words,

> Here the individual concept and the individual intuition do not stand in contrast to one another; it is rather a question of harmonizing the *function* of the understanding and that of beholding. The free play that is required concerns not representation but the powers of representation; not the results in which intuition and understanding are made concrete and in which they both come to rest, but rather the living excitation that occupies them.[49]

This interaction is pleasurable precisely because it enhances the interplay of those capacities that are the basis of all cognition. It serves the end of knowledge in general, and thence stimulates the total 'life feeling' of the judging subject.

The pleasure, which arises from this, has a distinctive character over and above disinterestedness. It has a claim to universal validity. Now in Enlightenment aesthetics (notably the French Classicists, Gottsched, the Swiss Critics and Baumgarten) any claim that the aesthetic judgment has to universality, must be demonstrable in terms of the judgment's objective content. Kant, however, brings about an extraordinary change of emphasis. For him, the universal validity of the aesthetic judgment is grounded on its origins in the cognitive conditions, which make objectivity and the self-possible. In questions of what *is agreeable* to the senses, each person is allowed their personal opinion. The enjoyment of sensible aesthetic configurations in contrast, is grounded on factors, which are essential to any cognition. Cassirer brings out what is involved here in a brilliant passage.

> Universality is not prevented by the individuality of the subject, for as it is true that the subjects have their life not only in passive sensory perceptions or in pathological desires, but can stimulate themselves to the free play of the representational powers, it is equally true that in such activity they employ one and the same essential basic function. In this functioning, which first properly makes the self into a self, every

> ego is akin to every other, and hence the function may be assumed in each other. The artistic feeling remains a feeling of self, but precisely as such it is at the same time a universal feeling of the world and life. The "self" detaches itself from its individuality when it objectifies itself in a construction of objective fantasy; its individual, unique stimulation is nevertheless not destroyed in this construction, but rather dwells powerfully in it and is communicated to all those who are capable of grasping it. Thus the subject is placed in a universal medium.[50]

The aesthetic judgment, then, has claim to universality because it is grounded in features that are essential to cognition. Here Cassirer finds the kind of insights which we saw to be basic to those links between imagination, aesthetics, and the self which Crowther eventually develops much further. However, in this immediate context, Kant is concerned more with the claim to intersubjective validity of the aesthetic judgment itself. This does not mean that in pronouncing something to be beautiful other people will necessarily assent to the judgment. Rather it means that (if our judgment is genuinely and purely aesthetic) they ought to, because it is grounded on a 'mutual accord' of the faculties, which is the basis of cognition in general. It is grounded in a rule which demands the concurrence of others but which is not simply a case of recognition in a definite concept.

Now throughout his exposition of Kant's doctrine of beauty, Cassirer assumes that this doctrine is applicable to works of art and natural beauty in equal measure. Having expounded the universal non-conceptual validity of the pure aesthetic judgment in general, Cassirer accordingly makes a leap from this to the condition of artistic creativity. In his eyes, this is a most momentous transition. For he takes it to be testimony to the fact that Kant makes it also, and in so doing, has, by a new route and different theoretical framework,

> arrived at the principal question standing at the crossroads of all aesthetic discussions in the eighteenth-century by a new route and in a completely different systematic connection.[51]

This question concerns the relation between rule and genius. For Kant, genius—like the universality of the pure aesthetic judgment—cannot be established by a conceptual procedure. Rather it is an innate natural gift, which produces exemplary rules for the creation of fine art. It is particular employment of reason, which is grounded in the creative spontaneity of the individual. As Cassirer observes,

> Genius and its act stand at the point where supreme individuality and supreme universality, freedom and necessity, pure creation and pure lawfulness indissoluble coalesce. In every line of its activity it is thoroughly original but nonetheless thoroughly exemplary. For just where we stand in the true focus of personality, where the latter gives itself purely without any external consideration and expresses itself in the

> individually necessary law of its creating, all the accidental limitations clinging to the individual in his particular empirical existence and his particular empirical interests fall away. In its immersion in this unadornedly personal sphere genius finds the secret and the power of universal communicability, and each great work of art presents nothing but the objectification of this basic power.[52]

Cassirer, then, sees the product of artistic genius as an elevation of personal vision to a level of universal validity. Through expression in a rule—governed artistic medium, the genius is able to invest that medium with an aesthetic significance, which it did not possess before. But because this involves an adaptation and reconfiguration of a publicly accessible medium, the artist's individual achievement is exemplary, i.e., can act as a model for other artists to take up and refine.

The effect of this doctrine

> [s]ignifies the achievement of a reconciliation between two diverse spiritual worlds, for it shares a crucial motive with the fundamental outlook of the Enlightenment, while on the other hand shatters the conceptual schema of the philosophy of Enlightenment from within.[53]

This is probably why Cassirer did not end his analysis of Enlightenment aesthetics with Kant. On the one hand he is picking up the theme of artistic genius as a formative power which he previously traced in Shaftesbury and others, but on the other hand his articulation of genius thereafter forms a point of departure for Romantic art theorists such as Schelling and Friedrich Schlegel. However, Kant is more than an instigator of Romantic aesthetics, for he reconciles this viewpoint with the Enlightenment advocacy of reason. For him, art and the aesthetic are not the exclusive, unifying characteristic of consciousness, rather they are one mode of unification which connects with other such modes but which (like the other modes) preserve their logical distinctiveness. Kant affirms this fact by emphasizing, for example, that whilst art has a rational core the fact that this is grounded in genius means that it is logically distinct from scientific creativity. As Cassirer says,

> This decisive difference lies solely everything which pretends to be scientific insight, as soon as it is to be communicated and established, possesses no form for this save that of the objective concept and objective deduction. The personality of the creator must be expunged if the accuracy of the result is to be protected. Only in the great artist is this division non-existent, for everything he gives is endowed with its peculiar and supreme value only through what he is.[54]

Cassirer identifies a similar affinity and difference in Kant's treatment of the sublime. In Part II of this chapter we noted that Cassirer's reading of Burke's theory of the sublime as centering on transcendence is one which transposes

Kant's position on to Burke. This becomes clear when Cassirer expands Kant's theory. Kant emphasises the fact that when presented with vast or mighty natural objects, our very inability to comprehend or resist such phenomena in physical terms serves to illuminate our capacity to comprehend or resist them through reason or the affirmation of our moral vocation. Here aesthetic contemplation and moral awareness are closely entwined, but Kant manages nevertheless, to preserve the logical distinctiveness of the former.

In noting this, Cassirer observes that:

> Our intuition remains aesthetic only when it views the self-determination of our mental faculties not in and for themselves but equally through the medium of the intuition of nature; when it reflects the inner in the outer and vice versa. Such a mutual mirroring of the ego and the world, of feeling of self and feeling of nature, comprises for both of us the essence of aesthetic contemplation in general and also the essence of contemplation which finds its expression in the sublime.[55]

The link between nature and our sense of self as free and rational is also found in art, by virtue of the significance of genius. The realm of nature construed in scientific and mechanistic terms seems to defeat any attempt to find an element of moral purpose in it. The two major elements of our existence—as both natural and free moral beings—seem to be in conflict with one another. Cassirer, however, notes that the Kantian aesthetic is able to offer reconciliation. In his words,

> Even if the objective agreement of nature and freedom remains a never-completed task, even if the paths of the two only intersect at infinity, their full subjective unity is actualized within the sphere of concrete consciousness itself, in the feeling of art and the creating of art. Hence, in the free play of the powers of the mind, nature appears to us as if it were a work of freedom, as if it were shaped in accordance with an indwelling finality and were formed from the inside out—while on the other hand the free creation, the work of artistic genius, delights us as something necessary and therefore as a creation of nature.[56]

Having elaborated Cassirer's reading of Kant's aesthetics, we will now offer a critical interpretation of it.

IV Issues Arising from Cassirer's Interpretation of Kant's Aesthetics

In the first two sections of this chapter we considered Cassirer's discussion of Enlightenment aesthetics. It was shown how Cassirer identifies four aspects that are requisite for the formation of a fully developed philosophical aesthetics. These are—the objectivity of beauty, the subjective response to it, art as a formative principle, and comprehension of all these factors in a systematic framework. All these factors are brought into a provisional alignment by Baumgarten. Cassirer's exposition of Kant indicates the way in which these are further integrated in a highly original way by means of the Critical aesthetic theory.

The formative basis of aesthetic experience is shown to be an active principle—a subjective variety of teleological judgment. Cassirer's subtle analysis of this is one, which links the teleological judgment *per se* to the constitutive level of experience as well as to the regulative. The aesthetic appears, accordingly, as a distinctive mode of a more general act of spiritual (in the sense of meaning—giving) formation. This, as we have shown in previous chapters is, for Cassirer, the basis of symbolic form per se. Any such form is a logically distinctive employment of the basic symbolic function.

Perhaps the most remarkable aspect of Kant's achievement is the way in which he rearticulates the objective basis of the aesthetic judgment. Rather than grounding this in concepts, he relates it to those subjective conditions, which make objective judgment possible. Objectivity in aesthetics is restated as a claim to universal validity based on the *intersubjective* conditions of communication. In making this switch Kant at the same time elucidates the nature of subjective aspect of aesthetic judgment. It is a feeling of pleasure, which arises through the interplay of those same factors (imagination and understanding), which are the basis of all cognition, *and* the aesthetic judgments claim to universal validity.

For Cassirer, then, Kant gives us a systematic philosophy of the aesthetic, which, in evidencing its distinctiveness, its relationship with other forms of knowledge, and its relation to general teleology, establishes the foundations of art as a symbolic form.

Two questions arise from all of this. First, is Cassirer's reading of Kant a correct interpretation, and second can Kant's aesthetic theory sustain the burden of significance which Cassirer places upon it, i.e., is it really as philosophically sound as Cassirer believes? The first question can be given consideration now, but the second one can be dealt with adequately only after we have discussed Cassirer's major account (in the *Essay on Man*) of art as symbolic form.

Speaking generally, Cassirer's reading of Kant is original and viable. However, a number of key difficulties can be identified (which will also re-emerge in the *Essay on Man)* The most central of these has already been illuminated—when we described Cassirer as reading Kant's general aesthetic as having equal applicability to both nature and art. This is more difficult than Cassirer presents it to us. Interestingly Kant himself never links the notion of disinterestedness to art and with some good reason. The relationship between imagination and understanding is necessarily different in the aesthetic appreciation of art, from the way in which it is the aesthetic appreciation of nature. Given some natural form we can enjoy its appearance without knowing what kind of thing it is. However, if we are to appreciate *artistic* beauty it is logically presupposed that we believe it to be a work of art. Here a definite concept of the object plays a necessary mediating role. This tells us that imagination and understanding have a less playful relation than in the aesthetic enjoyment of nature.

The suggestion is confirmed by a feature of fine art that Cassirer does not even consider—its status as 'aesthetic idea'. Kant defines this as '[t]hat representation of the imagination which induces much thought, yet without the possibility of any definite thought whatever, i.e., *concept,* being adequate to it, and which language, consequently can never get quite on level terms with'.[57] The relationship here is not the general free play of imagination and understanding that is found in the pure aesthetic judgment. The reason why is spelt out by Kant in a subsequent passage:

> If, we attach to a concept a representation of the imagination belonging to its presentation, but inducing solely on its own account such a wealth of thought as would never admit of comprehension in a definite concept, and as a consequence, giving aesthetically an unbounded expansion to the concept itself, then the imagination here displays a creative activity, and it puts the faculty of intellectual ideas (reason) into motion.[58]

For Kant, understanding is a capacity to set limits, but reason in general is the striving to comprehend totality and extend limits. In this passage Kant makes it clear that in the aesthetic idea imagination is in harmony with the latter, not the former. What we have is an instance of illumination not play. The concept involved is clarified and its scope made vivid by its mode and sensible presentation in an aesthetic idea.[59]

The problematic link of art and the aesthetic also recurs (eventually) in Cassirer's treatment of Kant's affirmation of the universal validity of the pure aesthetic judgment. Cassirer seems to regard Kant as having satisfactorily solved the problem of universal validity. However, there is an important gap in his reading. If the pure aesthetic judgment is, as Kant argues, grounded in the harmony of two capacities which are essential to any cognition whatsoever, then this shows that such a mode of judging has a universal (in the sense of accruing to be human *per se)* significance. But this does not *ipso facto* entitle us to demand the assent of others to our judging a form to be aesthetically meritorious. It may be that all aesthetic form engages imagination and understanding but this harmony may be brought about in different ways by different works. Some people will be sensitive to some such combinations but not to others. The aesthetic capacity may be universal, but the question of which specific forms engage it and to what degree is contingent on the particular experience and sensitivity of the individual. Kant himself endeavours a negotiation of this problem as follows (he is talking about the harmony of imagination and understanding): 'but this disposition of the cognitive powers has a relative proportion differing with the diversity of objects that are given. However, there must be one in which this internal reaction suitable for quickening (one faculty by the other) is best adapted for both mental powers in respect of cognition (of given objects) generally'.[60]

Now, as Crowther has shown in some detail, the assumption that there is one proportion of faculties best suited to stimulating their harmonious interaction is only an assumption.[61] True there must be a general likeness in how the faculties interact in order for one human being to be able to communicate with another. But this logical presupposition does not—without further extensive argument—entail that what gives the harmony of the faculties its optimum stimulation must also be alike in all people. Unless this can be established, we have no grounds for demanding the assent of others for a particular aesthetic judgment. The claim to universal validity in other words stops short of the level of individual judgment.

Cassirer, unfortunately, does not adequately consider the real issues involved in Kant's 'deduction' of the universality of the pure aesthetic judgment. In fact, he seems to make a claim which Kant himself does not. It is at this point that the problematic of the general art/aesthetic relation re-emerges. Cassirer (as was shown earlier) regards the product of artistic genius as the 'objectification' of the power of 'universal communicability'. But Kant himself does not say this, and again with very good reason. Cassirer's problem is that whereas he gives great significance to the concrete art object in his discussion of the Enlightenment aesthetics he pays little attention to it in the context of Kant. Instead he is preoccupied by its subjective aspect—the nature of genius.

Here it is important to note that Kant himself gives due emphasis to genius but not at the expense of the objective being of fine art. Cassirer barely even considers what Kant declares to be the 'primary property' of genius, namely *originality*.[62] To emphasize originality is to emphasize not only the fact that products of genius are exemplary (i.e., can serve as models for others) but that such works are exemplary precisely because they *differ* from other works in positive ways which offer new ways of working within a particular art medium. To recognize difference in the context logically presupposes a *context of comparison*.

Put generally this means that what makes a work into a work of fine art is its breaking with an established tradition. Aesthetic merit in the case of art, in other words, is not sufficiently explainable in terms of the universality of the harmony of the faculties. It necessarily involves a dimension of empirical, or more precisely, historical locatedness, whereby its innovations can be recognized. The universal validity of a work of fine art will be based (although Kant himself does not emphasize the fact) on its position in a historical horizon as well as upon formal consideration.

Overall, then Cassirer's account of Enlightenment aesthetics has shortcomings, but effectively discloses how, in that era, aesthetics as a discipline is progressively articulated in a way that links the aesthetic to fundamental cognitive concepts and relations. We have basic features of aesthetic cognition laid out before us. In the important chapter on art in *An Essay on Man*, Cassirer continues this account, and also formulates the outline of a theory of art of his own. This theory, as we shall see, allows links to Crowther's aesthetics to be made. These resolve his difficulties concerning the exact relation between art and aesthetic judgment *per se*, and of the nature of universal validity in aesthetic judgment.

Notes

1 The best account of Cassirer's interpretation of Enlightenment aesthetics is Marion Laushke, *Ästhetik im Zeichen des Menschen: Die ästhetische Vorgeschichte der Symbolphilosophie Ernst Cassirers und die symbolische Form der Kunst* (Hamburg: Felix Meiner, 2007).
2 Ernst Cassirer, *The Philosophy of the Enlightenment*, ed. James P. Pettegrove, trans. Fritz C.A. Koelin (Princeton: Princeton University Press, 1968), 276–277.
3 Ibid., pp. 293–294.
4 Ibid., p. 297.
5 Ibid., p. 298.
6 Ibid., p. 300.
7 Ibid., p. 303.
8 Ibid., pp. 306–307.
9 Ibid., pp. 308–309.
10 Ibid., p. 310.
11 Ibid., p. 311.
12 Ibid., p. 314.
13 Ibid., p. 317.
14 Ibid., p. 321.
15 Ibid., p. 323.
16 Ibid.
17 Ibid., p. 326.
18 Ibid., p. 327.
19 Ibid.
20 Ibid., p. 330.
21 Ibid., p. 329.
22 Ibid., p. 331.
23 Ibid., p. 333.
24 Ibid., p. 336.
25 Ibid.
26 Ibid., p. 338.
27 Ibid., p. 339.
28 Ibid., p. 341.
29 Ibid., p. 343.
30 Ibid., p. 348.
31 Ibid., pp. 348–349.
32 Ibid., pp. 352–353.
33 Ibid., p. 354.
34 Ibid., p. 329.
35 Ibid.
36 Edmund Burke, *A Philosophical Enquiry into the Origin of Our Ideas of the Sublime and Beautiful* (Oxford: Oxford University Press, Oxford, 2008), 136.

37 Curiously in this analysis Burke could be accused of the same failing which Cassirer assigns to the French classicists. For the need to have the "finer parts" stimulated other than by honest labour is surely the prerogative of a leisured class in a society which is prone to boredom. Burke, in other words, unacceptably elevates a historically specific susceptibility into a universal truth.
38 Crowther, *AE*, p. 20.
39 The fact that Cassirer's position on Kant was worked out well before his discussion of Enlightenment aesthetics explains why his analysis of the Enlightenment seems always to be tending towards Kant as its culmination. However, I will show that there are good logical reasons for such a narrative.
40 Ernst Cassirer, *Kant's Life and Thought*, trans. James Haden (London and New Haven: Yale University Press, 1983), 279–280.
41 Ibid., p. 288.
42 Immanuel Kant, *The Critique of Judgment*, trans. J.C. Meredith (Oxford: Oxford University Press, 1973), Book II, 32–33.
43 Cassirer, *Kant*, p. 288.
44 Ibid., pp. 299–300.
45 Ibid., p. 278.
46 Ibid., p. 307.
47 Ibid., pp. 309–310.
48 Ibid., p. 312.
49 Ibid. pp. 315–316.
50 Ibid., p. 319.
51 Ibid., p. 320.
52 Ibid., pp. 321–322.
53 Ibid., p. 323.
54 Ibid., pp. 322–323.
55 Ibid., p. 330.
56 Ibid., p. 333.
57 Kant, *The Critique of Judgment*, p. 176.
58 Ibid., p. 177.
59 For an approach which develops this position in relation to the sublime, see Chapter 7 of Crowther's *KS*.
60 Kant, *The Critique of Judgment*, p. 83.
61 See Crowther, *TKA*, pp. 93–96.
62 See Kant, *The Critique of Judgment*, p. 168.

4 Cassirer's Theory of Art and Crowther's Aesthetics

Cassirer published a few early papers on aesthetics, the most important of which is 'Mythic, Aesthetic, and Theoretical Space' of 1931.[1] Here he makes the point that art involves the 'creation of individual images', 'the medium of pure form' and 'the sphere of pure representation'. These are important concepts, but Cassirer does little more than affirm the fact of their importance, and link them to some important historical figures. It is only in the dedicated chapter on art in *An Essay on Man* of 1943 that he offers a more substantial discussion of these thematics.

His chapter is divided into three parts. The first one starts with some general points about the history of aesthetics, and then some further general points about art as a formative activity. Cassirer's second part concentrates on the significance of the relation that exists between art and imagination; and his third part refutes some rival theories to his own. We will look at each part in turn identifying weaknesses and endeavouring, in due course, to rectify them through a deployment of Crowther's aesthetics.

I The History of Aesthetics and Objective–Subjective Emphases

Cassirer begins by, in effect, taking up the story of aesthetics at the point at which he left it in his analyses of Enlightenment aesthetics and Kant's aesthetic theory. He affirms that whilst beauty is one of the most 'clearly known of human phenomena' its overlap with theoretical or practical functions has tended to prevent an adequate philosophical understanding of its nature.[2] Baumgarten's aesthetics offers a partial basis for understanding the distinctiveness of art and beauty, and in Kant's aesthetics this distinctiveness is fully illuminated.

As we have already seen, Cassirer argues that the great structural issue which all aesthetic theory must negotiate is the fact that the phenomenon of art has both objective and subjective poles. Theorists tend to veer towards one pole rather than the other in their analyses. Cassirer allows that in different contexts one or the other can be emphasized. This is in close keeping with a fundamental of his epistemology—the reciprocal dependence of body and mind. As he says in *The Logic of the Cultural Sciences*,

'The "reality" that we apprehend in perception and direct intuition gives itself to us as a whole in which there are no abrupt separations. And yet it is both "one and twofold"; for we apprehend it on the one hand as the reality of a thing and on the reality of a "person". We need physical science to explain the thing aspect, and concepts of form and Style from the "person".'[3]

Cassirer's approach in *An Essay on Man* keeps the thing aspect and "person" aspect very close together, emphasizing, for example, how individual style determines the artwork's physical (or imaginatively intended) appearance, and also the way beauty (and other 'subjective' aspects of the experience of arts) are closely correlated with the compositional unity of the artwork. He spends the rest of this section by proceeding from consideration of issues connected with the objective aspect, to issues that put more emphasis on the subjective one.

The focus of his analysis of the objective aspect is the notion of imitation. Imitation theories of art from Aristotle to the Classicist theorists of the Enlightenment do not see art as imitative in the sense of merely reproducing a given reality, for they also take account of the creative contribution of the artist. This contribution entails that what is given in the artwork is a kind of rule-governed deviation from reality. However, a problem is thus created. In Cassirer's words,

> how can we improve on our model without disfiguring it? How can we transcend the reality of things without trespassing against the laws of truth? From the point of view of this theory poetry and art in general can never be anything but an agreeable falsity.[4]

The inadequacy of the notion of imitation is, for Cassirer, also highlighted in historical terms through a Rousseau-inspired late eighteenth-century cultural tendency, which sees art as a vehicle primarily for the expression of the artist's emotions. What Cassirer has in mind here is art of the *Stürm und Drang* type—a point that is showed by his use of a momentous quotation from the young Goethe.

> Art is formative long before it is beautiful, and it is still true and great art, indeed often truer and greater than "fine art" itself. For in man's nature there is a will to create form which becomes active the moment his survival is assured. As soon as he does not need to worry or to fear, like a demi-god, busy, even in his relaxation, he casts around for a material into which he can breathe his spirit.[5]

This quotation also presents Cassirer's own theoretical position in outline. The notion of imitation is not an adequate expression of the objective dimension of art, because it does not highlight arts formative impetus and its deep expression of the spirit. Cassirer is however, very careful to keep this latter emotional aspect in its proper place. He notes how Croce and Collingwood drastically overplay it by misconstruing 'the whole constructive process which is a prerequisite both of the production and contemplation of art'.[6] Actually, Cassirer's treatment of Croce and Collingwood is rather unfair in its cursoriness, but it is certainly true that both of them radically downplay the significance of art as a mode of giving form to sensible material objects. This for Cassirer is the central formative feature from which arts distinctive emotional effects follow on.

He now makes his most important move by saying of art that 'It is not an imitation but a discovery of reality'.[7] Indeed,

> We must classify our sense perceptions and bring them under general notions and general rules in order to give them an objective meaning. Such classification is the result of a persistent effort toward simplification. The work of art in like manner implies such an act of condensation and concentration.[8]

Beauty and theoretical truth both center upon the unification of a manifold but whilst 'language and science are abbreviations of reality; art is an intensification of reality. Language and science depend upon one and the same process of abstraction; art may be described as a continuous process of concretion'.[9] The work of art does not address the physical qualities or causes of things but rather provides an 'intuition' of their *forms*. Hence, 'the artist is just as much a discoverer of the forms of nature as the scientist is a discoverer of facts or natural laws'.[10]

Cassirer's reasoning on these matters gravitates around his sense of what might be called the *superabundance* of phenomenal aspects. Even the simplest sensible item can be perceived from many different positions or under different lighting conditions. This means that its basic visual form or structure is veiled under the superabundance of its different aspects or appearances. The artist lifts this veil and reveals the 'realm of pure forms'.[11] Cassirer shows us the example of the landscape painter where

> what he gives us is the individual and momentary physiognomy of the landscape. He wishes to express the atmosphere of things, the play of light and shadow. A landscape is not "the same" in early twilight, in midday heat, or on a rainy or sunny day. Our aesthetic perception exhibits a much greater variety and belongs to a much more complex order than our ordinary sense perception. In sense perception we are content with apprehending the common and constant features of the objects of surroundings. Aesthetic experience is incomparably richer. It is pregnant with infinite possibilities which remain unrealized in ordinary sense experience. In the work of the artist these possibilities become actualities; they are brought into the open and take on a definite shape.[12]

Cassirer's use of the term 'pure forms' in relation to the objects of the artist's investigations is a little misleading; it perhaps implies an objectivity of structure that is too akin, say, to psychological investigations of the conditions of visual perception. However, what the immediate foregoing quotation is pointing towards is that the visual artist uncovers perceptual aspects, which sustain the (as it were) shifts macro-appearances of visible things. As Crowther points out:

> Normally when we see a mountain we do so without noticing the relationships of light, texture, and colour which constitute its natural appearance. However, the painter composes his image in a way which thrusts such qualities to the forefront of our attention. Our interest in the "mountain" as such is suspended in favour of appreciating the sheer structure of its appearance.[13]

In addressing the fabric of visual appearance the artist generates a corresponding image, but one which is highly selective and stylized. This gives a point of intersection between art's objective and subjective aspects. In Cassirer's words,

> The artist chooses a certain aspect of reality, but this process of selection is at the same time a process of objectification. Once we have entered into this perspective we are forced to look at the world through his eyes. It would seem as if we had never before seen the world in this particular light. Yet we are convinced that this light is not a momentary flash. By virtue of the work of art it has become durable and permanent. Once reality has been disclosed to us in this particular way, we continue to see it in this shape.[14]

It is worth giving emphasis to the fact that Cassirer extends this approach, also to other art forms. Authors such as Euripides, Shakespeare, Cervantes, and Moliere, for example, do not, according to Cassirer, entertain us

> with detached scenes from the spectacle of life. Taken in themselves these scenes are but fugitive shadows. But suddenly we begin to see behind these shadows and to envisage a new reality. Through his characters and actions the comic and tragic poet reveals his view of human life as a whole, of its greatness and weakness, its sublimity and absurdity.[15]

In these observations Cassirer is meaning clearly how plot or narrative weave episodes into a unified whole which illuminates specific aspects of the human condition. However, it is also consistent with Cassirer's approach to see this illumination as extending to a deeper level in terms of the scope of language itself.

Again Crowther is useful here. He describes the unity of plot and narrative as a case of 'superstructural' formal qualities. But there is also an 'infrastructural' level where the artistic articulation language and indeed musical form parallels the revelatory possibilities of visual art noted above. In Crowther's words,

> Normally, for example, in reading or speech we are concerned with the sense of words and sentences, and not with the phenomenal qualities which make this sense possible. However, sometimes in ordinary language, and always in poetry and literature, we remark not just upon what is said, but upon how it is said. We appreciate those qualities of, say, homophony, rhythm, metre, or imagery, which are the phenomenal matrix of a 'text's' sense. Similarly, whilst much of our attention to sound or music is simply a case of being carried along by the beat or rhythm, there are occasions (and this is almost always so with serious music) when we remark upon and enjoy the patterns of tone, timbre, melody, and harmony that constitute a work's phenomenal fabric of sound.[16]

Before addressing Cassirer's consideration of the more subjective pole of the phenomenon of art, we must question something about the point of intersection between the objective and subjective poles. Cassirer declares confidently that

> when absorbed in the intuition of a great work of art we do not feel a separation between the subjective and objective worlds. We do not live in our plain common place reality of physical things, nor do we live within an individual sphere.[17]

Plastic, musical, and literary forms 'have a real universality'.[18] Cassirer is surely right on these points. There is something about the artwork which adds to our knowledge of the world but which is achieved through the personal style of the artist. At this point, however, we hit a problem. Cassirer needs to clarify his criterion of universality and objectivity in formulating this position. The most he offers to us is the following:

> If the work of art were nothing but the freak and frenzy of an individual artist it would not possess this universal communicability. The imagination of the artist does not arbitrarily invent the forms of things. It shows us these forms in their true shape, making them visible and recognisable.[19]

The term 'universal communicability' is the tell-tale phrase here. Cassirer invokes Kant's deduction of the universality of aesthetic judgments, which are directed towards nature, and applies them—without further ado—as the basis of the artwork's objective intelligibility. This is a wrong move for several reasons. The first is that (as we showed in the discussion of Cassirer's understanding of Kant's aesthetic theory in the previous chapter) judgments about the universal validity of aesthetic judgment of art require very different consideration from those which Kant shows in relation to aesthetic judgments of nature.

The second reason why Cassirer is misguided is that Kant's problem of universal communicability is about questions of evaluation and not questions of description. Surely if an artwork is not just the 'freak and frenzy of an individual artist' it must have an *objective legibility* which is determined by it being inscribed in a publicly accessible semantico-syntactic code. To understand in the crudest sense what it is 'about' requires that we understand what it is a picture of, or what it is a poem or novel about, or what range of emotional resonance a piece of music is invoking. Truly this does involve an issue of 'universal communicability', but this is in no part the same normative notion as that invoked by Kant in relation to pure aesthetic judgments.

Cassirer's main problem here is that he has no real sense of the notion of *iterability* (a term which we will return to in more detail, a little further on). Artworks can be objectively intelligible only if the signs, or more accurately, representations codes which they use, involve *rules* which allow the pictorial (or whatever) sign to be recognized in whatever context it occurs, and in the absence of the kind, or specific, thing or state of affairs to which it is referring. It is this iterability, which makes the work of art into more than, as it were, an exclamation of purely personal significance. Of course, this is not all that there is to the artwork because, with art, we are not interested just in what is represented but also (as Cassirer does recognize) in *how* it is represented, i.e., the artist's individual expression. Cassirer notes the latter, but he does not give us an adequate theory of the former. Both are necessary if we are to have an adequate concept of 'art'.

The fact that Cassirer does not offer criteria of iterability for the different art media is rather odd. The reason for this is that the nature of the sign and the semantico-syntactic code in which its functions are what the idea of *symbolic form* is actually all about. Obviously, this is a topic which will be returned to as our argument develops.

Before doing that, it is important to consider Cassirer's points about how the objective and subjective poles of art are bridged in our aesthetic *response* to the artwork. On this topic, Cassirer criticizes Plato and Tolstoy's ways of linking art and emotion. Yes, we do respond to art with passion, but this is not just the same sort of passion, which we feel in everyday existence. In the words of Cassirer 'the image of a passion is not the passion itself. The poet who represents a passion does not infect us with a passion'.[20]

This is a really important point. If art is to be of profound significance, the feelings that it conveys to us must relate to life, but not be *exactly* the same as those that we experience in real life situations. But what *is* the difference? Cassirer notes that the poet gives the 'image' of passion, but what does this actually mean? He offers us no further illumination of the issue. Instead he focuses on the psychological dimension of what is involved. In encountering art, we leave the pressures and compulsions of real life emotions behind, but the responses, which we have to the artwork, are, in a sense stronger precisely because of their relative independence from real life. There is an 'aesthetic freedom' involved which

> means that our emotional life acquires its greatest strength, and that in this very strength it changes its form. For here we no longer live in the immediate reality of things but in a world of pure sensuous forms. . . . The passions themselves are relieved of their material burden. We feel their form and their life but not their encumbrance.[21]

Furthermore,

> what we feel in art is not a simple or single emotional quality. It is the dynamic process of life itself—the continuous oscillation between opposite poles, between joy and grief, hope and fear, exultation and despair. To give aesthetic form to our passions is to transform them into a free and active state. In the work of the artist the power of passion itself has been made a formative power.[22]

The phenomenology of aesthetic feeling, which Cassirer is describing here, has kinship with what in Chapter 2 we called the *act-association continuum*. This is the space of imaginings and affective intonations that constantly permeates our negotiations with life as well as our reflections upon them. It is in the fluctuating relation between this and our sense of self that the aesthetic narrative of self-consciousness described by Crowther is set up as a developing and creative whole.

However, in the foregoing passage Cassirer himself is describing the phenomenology of aesthetic feeling as it arises in relation to art. His description has *some* affinity with an approach which, in broader terms (as we have already indicated) he does not accept, namely the Croce-Collingwood theory. The essence of the Croce-Collingwood theory is to emphasize how, in experiencing art, the work's audience must re-create the inspirational process which led the artist to create his work. Similarly, Cassirer tells us that

> we cannot understand a work of art without, to a certain degree, repeating and reconstructing the creative process by which it has come into being. By the nature of this creative process the passions themselves are turned into actions . . . art turns all these pains and outrages, these cruelties and atrocities into a means of self-liberation, thus giving us an inner freedom which cannot be attained in any other way.[23]

Now whilst Croce-Collingwood approach diminishes the significance of the physically embodied artwork, Cassirer—as we showed earlier—does not. But if this is the case Cassirer has really to do more than describe only the phenomenology of the aesthetic feeling. He has got to link this very firmly with those objective characteristics of the physically embodied artwork, which determine the distinctive characteristics of aesthetic feeling.

This is another reason why Cassirer must give emphasis to the *iterable* aspect of the artwork, i.e., its specific semantico-syntactic structure. However, even if he had offered us an adequate theory concerning this, it would not completely resolve the problem presently under consideration. The additional factor which needs clarifying is the role of the artist's individual style in shaping the particular iterable medium in which he is working.

For the moment, we are content to note major difficulties for Cassirer as they arise in the course of his exposition. They will be rectified further on.

To continue, then, having analysed the objective and subjective poles of art, Cassirer moves on to a consideration of the phenomenon of beauty. For him, beauty is not a property of things; it is a relational phenomenon with a deep connection to human cognitive activity. Cassirer affirms that:

> Beauty cannot be defined by its mere *percipi*, as "being perceived"; it must be defined in terms of an activity of the mind, of the function of perceiving and by a characteristic direction of this function. It does not consist of passive percepts; it is a mode, a process of perceptualization. But this process is not merely subjective in character; on the contrary, it is one of the conditions of our intuition of the objective world. The artistic eye is not a passive eye that receives and registers the impression of things. It is a constructive eye, and it is only by constructive acts that we can discover the beauty of natural things. The sense of beauty is the susceptibility to the dynamic life forms, and this life form cannot be apprehended except by a corresponding dynamic process in ourselves.[24]

At the start of this passage, Cassirer makes an extremely strong claim—that beauty is, in a sense, a condition of our having knowledge of an objective world. But *why* should it have this significance? The second part of the passage does not offer an explanation. In fact it looks to move in a slightly different direction, as though the perception of beauty is a 'discovery' of beauty and one, which requires a corresponding dynamic act of cognition. This would suggest something that Cassirer has already denied, namely that beauty is a property of things, although one, which requires a special capacity of mind in order to be recognized.

The point, which Cassirer is trying to make here, is perhaps explicable on the following lines. Earlier on Cassirer has linked art to a focusing of what we called the objective world's 'superabundance' of aspects. An embodied being can perceive even the simplest state of affairs in terms of many different perceptual aspects. Language enables us to 'abbreviate' this complexity and pick out specific aspects for closer and more sustained attention. It is this, which allows us to comprehend the superabundance of aspects in terms of the systematic and rule-governed relationships of an objective world.

Art plays a specific role here. It enables us to contemplate the dynamic structure of formal relationships and, thereby, to comprehend the nature of sensory presence of individual things or states of affairs with enhanced clarity.[25] The aesthetic enjoyment, which this contemplation involves, acts as a stimulus, or better, a *motive* for engaging in further cognitive exploration of this sort. The more such cognition is motivated, the clearer our conception of the dynamics of form becomes; and the clearer this becomes the more our sense of the objective structure of sensible presence is clarified. At the heart of such contemplative activity is a formative process. This is because the dynamics of form requires us to make complex perceptual connections, rather than simply recognize what kind of thing the sensible formation that we are attending to actually is.

Given this, Cassirer's description of the experience of beauty as a 'discovery' is quite an apt affirmation. It also links to Crowther's accounts of aesthetic exploration as an important factor in the ontogenesis of objective knowledge and the narrative of the self. An infant will be fascinated by the striking look of things and the way they fit together, but, as his or her cognitive orientation becomes more established, explorative attention will become gradually linked to only those things that stand out in some way—for example, by being beautiful.

Having raised the question of beauty, Cassirer now investigates its relation to art and nature. Some philosophers such as Croce (and indeed, Hegel and Clive Bell, although Cassirer does not mention them here) have claimed that natural beauty is enjoyed only insofar as it is analogous with artistic beauty. This might appear condescending to nature. Cassirer attempts to bypass this danger by distinguishing between 'organic beauty' and 'aesthetic beauty'. In his words:

> The organic beauty of a landscape is not the same as that aesthetic beauty which we feel in the works of the great landscape painters. Even when we, the spectators, are fully aware of this difference. I may walk through a landscape and feel its charms. I may enjoy the mildness of the air, the freshness of the meadows, the variety and cheerfulness of the colouring, and the fragrant odour of the flowers. But I may then experience a sudden frame of mind. Thereupon I see the landscape with the artist's eye—I begin to form a picture of it. I have now entered a new realm—the realm not of living things, I live now in the rhythm of spatial forms, in the harmony and contrast of colours, in the balance of light and shadow. In such absorption in the dynamic aspect of form consists the aesthetic experience.[26]

In this passage, Cassirer is taking a very Kantian approach. Kant makes a distinction between the 'agreeable' and the 'beautiful'.[27] The former is based on sensation (i.e., mere impacts on the senses) and the question of whether or not such a sensation if agreeable is entirely down to personal preference. The beautiful (as was shown in Chapter four) is not about mere sensations, but rather concerns relations between the formal elements in a sensible manifold. As was also shown, it has a claim to universal validity. The way Cassirer makes use of Kant's distinction is fairly clear. 'Organic beauty' parallels Kant's 'agreeable' and aesthetic beauty parallels Kant's formalist understanding of natural beauty. In the case of the latter, we go (in Cassirer's terms) beyond mere organic beauty when we appreciate nature *as if* it had—like art—an objective purposive structure.

This is plausible, but it is not a complete understanding of natural beauty. The reason why is that the agreeable has rather more aesthetic scope than either Kant or Cassirer give to it. In talking of our absorption in 'dynamic form' Cassirer means a relation to formal qualities of the sort which—following Crowther—we earlier called 'infrastructural'. However, we must also recall that Crowther posits a level of 'superstructural' formal qualities. He says, for example, that

> whilst in a late Cézanne *Mont St Victoire* one can enjoy the relationship between masses and textures of colour as such, one can also appreciate the relationship of the mountain with the landscape in a formal sense. One might, in this respect, enjoy the way the mountain sits (phenomenally speaking) upon the horizon, or seems to support the clouds. Indeed, one can marvel at the very fact that a cloudy day can be made to appear in such vibrant yet monumental terms.[28]

The superstructural qualities, which Crowther describes here in a picture of nature, can—surely—also be enjoyed in nature itself. All sorts of qualities such as the freshness of the wind, the fragrance of flowers, can go beyond what they are individually (i.e., merely agreeable relations) to compose a complex relational whole. Here we could enjoy the natural scene in purely aesthetic terms, although the parts from which the whole emerges were primarily agreeable. The phenomenon of wine tasting illustrates this even better. As a literal case, our enjoyment of wine's flavour should merely be agreeable. However, it is possible to enjoy the *subtleties* of its *relation* to our palate, and its comparative status in relation to other wines, which we have enjoyed.

> The point at issue here is not that Cassirer and Kant are wrong to logically distinguish aesthetic beauty from the realm of the agreeable but rather that there is an important boundary zone between the two. If Cassirer wishes to understand the phenomenon of beauty in the fullest sense, then he ought to take account of this dimension. Such recognition is one of the key insights in Crowther's account of the aesthetics of the self. We have practically significant experiences, and enjoy agreeable sensations for their own sake, but in the relation between all these an emergent relational pleasure or displeasure can arise in terms of the playful relation between these and other experiences and the overall narrative of the self.

II The Emergence of Imagination

Cassirer begins the next (much shorter) section of his chapter on art by affirming that all schools of aesthetic theory agree that art forms an independent 'universe of discourse'. The key to his confidence is the concept of *imagination.* We have already developed Cassirer's position by detailed reference to Crowther, but here Cassirer is concerned more to account for its historical emergence as a concept in aesthetics.

His first point of substance is to note how even in the aesthetics of classicism (in Boileau and Batteaux et al.) the notion of imagination is important, but only insofar as it is rigorously determined by rules determined by the search for Ideal form. However, from the mid-eighteenth-century (and especially in the work of the Swiss critics Bodmer and Breitinger) a change takes place.[29] The 'marvellous' and 'miraculous' are seen as, in Cassirer's words, 'the only subjects that admit of true poetical portraiture'.[30] Indeed, following Vico, Cassirer further declares that:

> Mankind could not begin with abstract thought or with a rational language. It had to pass through the symbolic era of myth and poetry. . . The poet and the maker of myth seem, indeed, to live in the same world. They are endowed with the same fundamental power, the power of personification. They cannot contemplate any object without giving to it an inner life and a personal shape.[31]

This suggests to Cassirer that romanticism and, especially Schiller's admiration for the Greeks is expressive of a yearning for a time when myth was real, and not merely an allegorical aspect of art. In Cassirer's terms, indeed, art never loses its connection with the 'divine' mythic past.

Having set a general historical concept for the emergence of imagination as an aesthetic concept, Cassirer now addresses its objective aspect in relation to the physical embodiment of art. In art images are 'externalized'.

> Externalisation means visible or tangible embodiment not simply in a particular material medium—in clay, bronze, or marble—but in sensuous forms, in rhythms, in colour pattern, in lines and design, in plastic shapes. It is the structure, the balance and order, of these forms, which affects us in the work of art. Every art has its own characteristic idiom, which is unmistakable and unexchangeable. The idioms of the various arts may be interconnected, as, for instance, when a lyric is set to music or a poem is illustrated; but they are not translatable into each other.[32]

However, if the idioms of the various art forms are not translatable into each other, why is it that we are entitled to regard them as all as expressions of a common factor, namely the image? Unfortunately, Cassirer does not offer any readily apparent criterion of 'image' as used in this context. This is why it is useful to have Crowther's theory of imagination in place as an explanatory tool. Cassirer, in effect, shows the concept of imagination has been progressively articulated in the history of aesthetics, and Crowther provides the most aesthetically relevant, contemporary account of it.

In the following two chapters, we will clarify in detail the image character of individual art media. However, in the first instance, it is useful to consider a useful general criterion of the artistic image provided by Susanne K. Langer. She notes that the term 'image' has become inescapably associated with the visual dimension.[33] To rectify this restrictiveness she revitalizes the term 'semblance'—from thinkers such as Jung and Schiller—in order to specify a general product of the imagination which is not tied specifically to the sense of sight. Langer generalizes thus:

> The setting forth of pure quality, or semblance, creates a new dimension, apart from the familiar world. That is its office. In this dimension, all artistic forms are conceived and presented. Since their substance is illusion or "Schein" they are, from the standpoint of practical reality, *mere* forms; they exist only for the sense or the imagination that perceives them—like the fata morgana, or the elaborate, improbable structure of events in our dreams. The function of "semblance" is to give forms a new embodiment in purely qualitative, unreal instances, setting them free from their normal embodiment in real things so that they may be recognized in their own right, and freely conceived and composed in the interest of the artist's ultimate aim—significance, or logical expression.[34]

Langer traces how the notion of semblance is given a different structure according to the nature of the different physical media that are involved. To take the two extremes of semblance—pictorial art on the one hand, and music on the other—Langer suggests that the former posits a 'virtual space' and the latter a 'virtual' time. In more detail she describes these virtualities as follows:

> The harmoniously organised space in a picture is not experiential space, known by sight and touch, by free motion and restraint, far and near sounds, voices lost or re-echoed. It is an entirely visual affair; for touch and hearing and muscular action it does not exist. For them there is a flat canvas, relatively small, or a cool blank wall, where for the eye there is as deep space full of shapes. This purely visual space is an illusion, for our sensory experiences do not agree on it in their report. Pictorial space is not only organized by means of colour (including black and white and the gamut of rays between them), it is created; without the organizing shapes it is simply not there. Like the space "behind" the surface of a mirror, it is what the physicists call "virtual space"—an intangible image.[35]

It is important to emphasize that Langer's notion of virtual space here has application to non-objective as much as conventional pictorial representation, in as much as any mark on a plane surface or identifiable edge to the plane serves to activate the surface. In the first case, the mark seems to penetrate the surface or sit on top of it, introducing thus the sign of a third dimension. In the latter case, insofar as the plane is manifestly edged (by a frame or whatever) it opens up the illusion of an empty space into which the viewer can move, of that of a full space, which would block such movement. In these cases, the third dimension is defined tacitly by reference to the spectator's possible movement.[36] Of music's virtuality, Langer tells us as follows:

> The semblance of this vital, experiential time is the primary illusion of music. All music creates an order of virtual time, in which its sonorous forms move in relation to each other—always and only to each other, for nothing else exists there. Virtual time is as separate from the sequence of actual happenings as virtual space from actual space. In the first place it is entirely perceptible, through the agency of a single sense—hearing. There is no supplementing of one sort of experience by another. This alone makes it something quite different from our "common sense" version of time, which is even more composite, heterogeneous, and fragmentary than our similar sense of space.[37]

The virtual time of music allows the experience of time to be monopolized by one sense, and comprehended accordingly in terms of its form and continuity—without reference to the distractions of everyday practical experience and its struggles.

By invoking Langer at this point, we are not only offering a viable assisting argument for Cassirer, but also one—in relation to virtual space at least—which has its partial origin in a common source, namely Adolf Hildebrand's *The Problem of Form in the Fine Arts*. Cassirer explicitly invokes Hildebrand to assist his arguments in that section of the *Essay on Man* which is being presently considered.[38] Langer makes use of Hildebrand even more extensively throughout her entire account of the relation between virtual space and pictorial representation.

Having, therefore, considered Cassirer's notion of the externalized image and the way it might be made more complete by reference to Langer, we now look at his interesting worries about how the notion of imagination has been misused.

These reservations center on German Romantic Idealism. In Fichte and (especially) Schelling, for example, imagination and art are elevated into having an *ultimate* philosophical significance. They see (although Cassirer does not give us much detail *how* they see it) the infinite as being the authentic subject matter of art. However, for Cassirer, this is unnecessarily restrictive in relation to the process of artistic creation. In fact, for him, nineteenth-century Realist Art actually leads 'to a more profound conception of artistic form'[39] which discerns important structural features of the forms of visual life and patterns of personal experience and social intercourse. At the same time, however, Realists tend to overreact against the Romantic standpoints by failing to understand adequately the *symbolic* nature of art.

From Cassirer's point of view, the artwork as image (or better, 'semblance' in Langer's sense) evokes aspects of experience, but it does not (contra Romanticism) touch some metaphysical substratum or (contra Realism) merely reproduced important aspects of the experiential world. Rather it seeks out fundamental visual and/or emotionally significant forms, which are *immanent* in experience. As Cassirer declares, the real subject matter of art

> is to be sought in certain fundamental structural elements of our sense experience itself—in lines, design, in architectural, musical forms. These elements are, so to speak, omnipresent. Free of all mystery, they are patent and unconcealed; they are visible, audible, and tangible.[40]

This fundamental level of form is not tied to any one kind of subject matter. There can be no hierarchies of, or excluded, genres in relation to the pursuit of it.

III Cassirer's Interpretation of Rival Aesthetic Theories

So far, then, in his chapter on art, Cassirer has looked at the history of aesthetics and the nature of art as a formative activity, and its origins in the imagination. In the final section of his chapter gives more detailed consideration to viewpoints in opposition to his own. He begins with 'psychological' theories. These for Cassirer have an advantage over metaphysical theories in that they do not have to provide a systematic definition of beauty. Basically they are concerned only with the 'fact' of beauty and a descriptive analysis of how it is experienced psychologically. But Cassirer affirms that such approaches cannot do much useful work, and tend to lose sight of important qualitative distinctions between beauty and other kinds of experience. They center overly on questions of degrees of pleasure.

Santayana's aesthetic is, for Cassirer, one example of such an approach. He notes that

> according to Santayana beauty is pleasure regarded as a quality of things; it is "pleasure objectified". But this is begging the question. For how can pleasure—the most subjective state of our mind—ever be objectified?[41]

Cassirer also notes Santayana's claim that 'Art is the response to the demand for entertainment'.[42] However, Cassirer is not in accordance with this claim either He holds that

> if this were the end of art we should be bound to say that art, in its highest achievements, fails to attend its real end.[43]

The point here is that great art often disturbs or unsettles as much as it elevates our feelings. Indeed as Cassirer says a bit later on, it requires powers of concentration in order to be appreciated, and this is surely at odds with Santayana's assignation of it to mere entertainment.

The reason why the enjoyment of art requires such concentration is that, for Cassirer (as we saw earlier), beauty is not an objective property of things, but a selective engagement which distinguishes key formal elements from the object's other aspects. Beauty is, in a sense, produced through this selective activity. As Cassirer affirms,

> To define beauty as "pleasure objectified" contains, therefore, the whole problem in a nutshell. Objectification is always a constructive process. The physical world—the world of constant things and qualities—is no mere bundle of sense data, nor is the world of art a bundle of feelings and emotions. The first depends upon acts of theoretical objectification, objectification by concepts and scientific constructs; the second upon formative acts of a different type, acts of contemplation.[44]

Unfortunately, Cassirer's critique of Santayana is not absolutely compelling. The notion of beauty as 'objectified pleasure' does not automatically exclude the idea of it involving some productive activity on the viewer's part. Cassirer's' refutational arguments need more substance here.

Having addressed Santayana, Cassirer now gives consideration to another psychological approach (of Romantic origin), which identifies art with a kind of waking dream. He finds a modern version of this in Bergson, whom he quotes as follows.

> In the process of art we shall find, in a weakened form, a refined and in some measure spiritualised version of the processes commonly used to induce the state of hypnosis.[45]

Cassirer is sceptical about this. The hypnosis analogy is wholly inappropriate because of the *passive* state that it centers upon. For Cassirer, as we have already seen, in contrast, the viewers engagement with art is highly active.

Cassirer makes a similar point in relation to Nietzsche's psychological understanding of art as a product of a state of spiritual intoxication (in *The Birth of Tragedy*). Against Nietzsche's view Cassirer declares that:

> Even in this theory of its psychological origin one of the essential features of art has disappeared. For artistic inspiration is not intoxication, artistic imagination is not dream or hallucination. Every great work of art is characterized by a deep structural unity. We cannot account for this unity by reducing it to different states which, like the Dream State and the state of intoxication, are entirely diffused and disorganized. We cannot integrate a structural whole out of amorphous elements.[46]

Unfortunately, Cassirer's critique is somewhat misleading in relation to both Bergson and Nietzsche. In Bergson's case, he centers on a more or less passing remark in an early work and overlooks the very complex links that Bergson makes between art and 'intuition' in his later works.[47] In relation to Nietzsche he completely overlooks the connections that are made between art and the formative activity throughout Nietzsche's writings.[48]

Indeed, whilst Cassirer emphasizes the 'deep structural unity' of the great artwork, this unity is primarily found at the level of form. Of itself this does not rule out some great artworks having their psychological origins in the kind of intoxication that Nietzsche describes. Indeed, because, for Cassirer, the art object serves to concentrate and intensify experience, one would have expected that the psychological states, which inspire it, are actually of a more diffuse and less controlled character.

Cassirer now goes on to give consideration to those theorists who posit a connection between art and play. He notes that, psychologically speaking, art and play bear a close connection with one another because both have a non-utilitarian significance which is unrelated to the realization of any practical ends. However, there are also some very important differences that Cassirer adduces. For example, in the phenomena of children's play, the play involves simulation which engrosses the participants to such a degree that (for a little while at least) no distinction is made between play and reality.[49]

In the artwork, in contrast, we take the semblance to be semblance—to be a vehicle for the expression of pure forms, which are immanent to reality, rather than just another aspect of reality. Indeed, it is the task of aesthetic judgment to distinguish between the genuine artwork and those that merely provide entertaining substitute realities of an escapist kind. It demands powers of concentration rather than a mere play of pleasurable feelings. And again, whilst children's play is in one sense a preparation for adult roles in life, fine art's function is neither a preparation of this sort nor the provision of pleasing diversions.

Cassirer's objections to the art/play connection so far, seem to be directed primarily at theorists such as Herbert Spencer. However, he also has Schiller in mind but in rather more sympathetic terms. Schiller assigns great importance to the link between art and 'play-drive' but he interprets 'play' in an extremely different way as a harmony of human sensible being and its higher cognitive capacities.[50] This, of course is a direct parallel to Kant's account of the aesthetic experience's playful harmony of imagination and understanding that was discussed at length in the previous chapter. Curiously, Cassirer does not make anything of this connection. Instead he notes another Kantian factor in Schiller's position as follows.

> Aesthetic contemplation and reflection, according to Schiller, is the first liberal attitude of man toward the universe. "Whereas desires seizes at once its object, reflection removes it to a distance and renders it inalienably her own by saving it from the greed of passion". It is precisely

> this "liberal", this conscious and reflective attitude which is lacking in child's play, and which marks the boundary line between play and art.[51]

Here, in, effect child's play is separated from art by the fact that the latter has an overt distance from, or disinterestedness in relation to, the demands of everyday life.

Again Cassirer's treatment of the art/play relation is very cursory and requires a much fuller treatment, especially in light of his sympathy with Schiller and Kant's own use of the notion of play in the aesthetic context, and, of course, in relation the important points we made about Crowther's emphasis on play and the aesthetic in the genesis of objective knowledge and the unity of self-consciousness.[52] Cassirer however, ends his chapter on art with an affirmation of this position overall. Art is a symbolic language, which is orientated towards factors other than those of an immediate practical or theoretical significance. It is concerned not with the uniformity of laws but with the multiformity and diversity of sensible and imaginative intuitions. In Cassirer's words,

> A mere duplicate of reality would always be of a very questionable value. Only as conceiving art as a very special direction, a new orientation, of our thoughts our imagination, and our feelings, can we comprehend its true meaning and function. The plastic arts make us see the sensible world in all its richness and multifariousness. . . . Poetry is, similarly, the revelation of our personal life. The infinite potentialities of which we had but a dim and obscure presentiment are brought to light by the lyric poet, by the novelist, and by the dramatist.[53]

Interestingly, Cassirer also makes some further points about art that are of vital importance, at the end of *An Essay on Man*. We defer consideration of these until after a broader critical discussion of the chapter addressed specifically to art.

IV A Critique of Cassirer, and Some Use Made of Ideas from Crowther

As we have just seen, the final section of Cassirer's chapter is extremely cursory in its treatment of alternative approaches to his own. This is not to say that his objections to them are wrong. They simply need to be made more substantial. Cassirer's problem here is one that recurs throughout his philosophy. He is too anxious to show the broader cultural and critical context in which symbolic forms are produced and interpreted, that this sometimes leads him to give a fairly restricted exposition of the symbolic form itself.

However, the basis of Cassirer's conception of art is at least clear. It is a mode of forming sensible or imaginative material in physically embodied images, which involve a selective presentation of pure forms. Using Crowther and Langer we have shown how this basic strategy can be bolstered by reference to the former's notion of infrastructural and superstructural formal qualities, and the latter's notions of semblance and virtuality.

What Cassirer offers us, in effect, is an account of how—from the Enlightenment aesthetics arose as a discipline through the joint progressive articulation of thinking about the arts and practice in the arts. He shows how the elements of what we have called aesthetic cognition (based on the enhanced reciprocity of object and subject of experience achieved through the creation and appreciation of art) are gradually understood as the basis of unique symbolic form. His own theory of art is a full expression of this recognition, and thence operates at the heart of those issues that are still centrally relevant to art and the broader life of art in culture.

However, in order to do justice to aesthetic cognition, we have to continue Cassirer's project, by negotiating some of its problems, and developing some of the ideas that are only hinted at in his presentations. For example, at the end of the last chapter we noted that in his treatment of Kant's aesthetics Cassirer does not satisfactorily resolve the relation between the aesthetic enjoyment of nature on the one hand and art on the other. Our enjoyment of the latter presupposes that we recognize a formal configuration *as* a work of art. In *An Essay on Man* Cassirer does not offer further clarification of this factor. In the main, he emphasizes how as a formative activity art unifies form in an active way, but at the same time he also affirms that our aesthetic relation to nature has a 'constructive' dimension. The only real difference between the two, which Cassirer shows, is found in his observation that in engaging with the artwork the viewer in some sense recreates the process of artistic creation itself. Cassirer unfortunately does not elaborate on this; neither does he begin to explain how the 'pure forms' in the artwork might guide such a response.

The main problem in all this is the way in which Cassirer (and to some degree Langer also) leaps from embodied semblance to the notion of pure form without taking account of a significant mediating factor. For to 'externalize' does not mean simply to embody in a physical medium, but to do so on the basis of a specific semantic-syntactic code. This is the factor of 'iterability' which we mentioned in Part One of the present Chapter. Cassirer touches on points involved in this notion on many occasions[54], but only in very general and opaque terms. Crowther's formulations of iterability are much more helpful. He defines it as follows:

> Language (and sign-systems generally) . . . have an iterable character, when their terms and grammatically correct constructions are intelligible without either the immediate presence of the things to which they refer, or knowledge of the intentions or contexts which inform their specific uses. The importance of this iterable dimension is that it is a necessary condition of any systematic signifying activity whatsoever. If

> signs were not iterable in this way they would have no third-person force. They would not be shareable. Meaning would be locked into a putative wholly private space of personal intentions.[55]

The importance of iterability is that it is the basis of systematic codification. If one knows the basic code in which a configuration of signs is inscribed—be it language or a specific linguistic format (such as fiction or poetry) or the scale systems of music, or pictorial representation—then such an inscription has intelligibility beyond the immediate context in which it was produced.

This enables *sharing* in a twofold sense. On the one hand, we can recognize what kind of thing is being represented in the particular inscription, and, on the other hand, we will be able to recognize personal inflections and style in the way it is inscribed, by virtue of its difference from other inscriptions of a broadly similar character. Both these factors of sharing are of vital importance in art. As Crowther goes on to say in relation to the visual arts: (although the point also admits of more general application),

> the iterable dimension gives a context of sameness wherein individual styles and mere repetitions can be clearly comprehended in relation to both their predecessors and successors. We value those who innovate or refine in relation to the code; they rejuvenate and change our sense of visual life. In such a context the temporal dimension is not just a mediating element (as is the case of natural forms) but neither is it the swampland of mere empirical history. Each work stakes a claim to attention on the basis of reciprocity between its appearance and that diachronic history of the code and medium which precedes its creation, and the subsequent history which its appearance helps create.[56]

Linking this to Cassirer it is clear that the artwork is a semblance of something not just by virtue of physical externalization, but also by virtue of being embodied in an iterable code. This means that even if we are attending to purely formal qualities we must be attentive to the fact, that, as inscriptions in a code, they have been generated through powers of human artifice, they have been brought forth in a shared public context as well as through the personal vision of the artist. The selectivity of the artist's revelation of pure form emerges not only in what he has gathered up and put into the work, but also through its relation to the tradition of the code—its points of likeness or difference in relation to other works, as well as its consonance or dissonance with our own personal experience of the world.

It is, then, the iterable dimension of artistic semblance and its various ramifications, which decisively separate our aesthetic experiences of art from those of nature at the objective level. Indeed, this point can be taken much further in terms of subjective responses. Cassirer holds (as we have shown at length) that our aesthetic response to art involve a distance (or in Kantian terms, disinterestedness). But in his terms so also does our aesthetic response to nature. In order to avoid contradiction Cassirer needs to articulate two different notions of disinterestedness, which manage to share something in common.

The problem is, however, that whilst in his discussion of Enlightenment aesthetics he emphasizes disinterestedness only secondarily—as an attitude prompted by the aesthetic object, in *An Essay on Man* he scarcely goes beyond this. He offers us a few more psychological details of the attitude, but does not focus adequately on how this attitude is shaped by the objective features of the particular kind of aesthetic object involved. The notion of pure image or semblance is not instructive in this context because natural beauty is enjoyed *as if* it were art (which is what Cassirer takes to be the case) then it too is semblance. The difference between artistic and natural disinterestedness cannot, accordingly be grounded on such a notion in itself. We need an extra consideration.

Again Crowther enables this problem to be resolved through a striking reinterpretation of what our criteria of disinterestedness in these contexts should be. Because Kant linked disinterestedness to contemplation, it has been universally assumed that it is a psychological attitude. Crowther's point, however, is that it is a *logical* characterization of aesthetic judgments and one, indeed, which has different varieties. In Crowther's words: 'if our appreciation of some phenomenon is founded solely on grounds which do not necessarily presuppose reference to its (in the most general sense) practical significance, then logically speaking, we would be entitled to characterize our experience of such a phenomenon as disinterested.'[57]

For Crowther, disinterestedness in this sense is the factor that is common to all aesthetic responses. However, the aesthetic enjoyment of natural beauty is disinterested in a *further* sense. In order to appreciate a beautiful natural object in purely formal terms, it is not presupposed that we know what kind of thing is the bearer of these properties. Our enjoyment requires no knowledge of the exact ontological status of the beautiful object, because it is grounded on the structure of its appearance alone. In this sense the enjoyment of natural beauty is *absolutely disinterested.*

In Crowther's terms we are able thus to make a contrast between the beauty of nature and the aesthetic enjoyment of art. Both involve disinterested pleasure in that we do not have to believe them to be of practical significance in order to enjoy their qualities of form. However, in contrast with our enjoyment of beautiful nature, in order to aesthetically appreciate art, it is logically presupposed that we know that the formal configuration *is* a work of art (i.e., that we recognize its ontological status as a thing which has been brought forth through powers of human artifice). This means that whilst our aesthetic enjoyment of nature is absolutely disinterested our aesthetic appreciation of art is more qualified; it is, in Crowther's terms *relatively* disinterested.[58]

However, this qualification has, itself, some positive consequences. The key to these is the concept of iterability. As we have argued, an inscription in an iterable code—be it linguistic, pictorial, or whatever—must be intelligible without the thing or state of affairs which it refers to being directly present to us, and without us having to have knowledge of the intentions which led the creator to formulate the inscription. To be iterable is to have a basic sense, which is recognizable beyond a signifying configuration's immediate context of production.

In the case of art, this fact is of major importance. In order to enjoy its distinctive presentation of form all that is presupposed is the ability to interpret it as an instance of a shared communicative code, to be able to recognize its stylistic differences from other instances of the code. What is *not* presupposed is any knowledge of the particular circumstances, which surrounded the works production and reception, or of biographical details of the artist's life. Often such knowledge will be available and it may in some ways add to our enjoyment, but the major point is that it is not a logical presupposition of it.

In this sense we might talk of iterability making a possible distinctive artistic notion of disinterestedness—a distinctive artistic mode of pleasure. This pleasure is not tied to the fulfilment of immediate practical concerns. Instead it feeds into what Crowther calls our 'global sense of life' namely 'our total immersion in the world as embodied subjects—that is, as beings whose rational, sensible, social, and historical existence is experienced fundamentally as an internally related and unified field.'[59] Through personal articulations of an iterable code, Crowther suggests that:

> We can identify with the specific artist's originality of vision, but without having to assimilate the minutiae of his or her life beyond the artwork. In art, the particular creator and the audience meet on common ground in an aesthetic mode of empathy which engages rational, sensible, and historical factors in an inseparable unity. Of course, other persons can tell us what interests them, what they value; and how they would imprint their presence on the fabric of things. But the sensuous particularity of art shows this with a directness which eludes the expressions of ordinary expressive language. When aesthetic form is energized by originality of articulation, it enables us to see something of the other from where he or she sees us—but on our own terms as well.[60]

These references to Crowther definitively resolve Cassirer's obscurities in relation to both the objective and subjective aspects of the aesthetics of art and nature. However, there are a number of other problems outstanding still. Earlier we showed how Cassirer affirms that the audience must re-create the inspirational process by which the artwork comes into being.[61] However, we also noted the fact that Cassirer offers no criterion whereby this re-creation can be linked to the objective features of the artwork itself. In the absence of this his position veers dangerously towards a point of view, which he officially rejects, namely the Croce-Collingwood affirmation of the primacy of the artist's inspirational states over the physically embodied art object.

As was also shown, Crowther's position emphasizes the opposite factor—the primacy of the artwork over the biographical context or specific inspirational states, which brought it forth. To make Cassirer consistent we must locate the process of the audience's re-creation away from *fantasies* of identification with the artist's actual experiential states, to the realisation of these in the finished physically embodied artwork. In one of his best passages (which we quote at length) Crowther shows how this can be done.

> All human beings die; but works of art need not. They preserve and declare ways of experiencing the world. This is not just a case of experience being translated into a more enduring form. Rather, it is at the same time, an overcoming of contingency by finite means. Each brushstroke, each stone each word, each note (together with the symbolic content which they bear) has its necessary place in the symbolic and necessary relation *vis-à-vis* the finished product of artistic labour. They are gathered up and redeemed. The individual artwork entails the existence of its unique creator, with his or her personal history. More than this, it presents that experience in a form which is directly accessible to other human beings. The artist's audience *inhabits* the work. It finds there the echo of its own existential problems, and ways of elucidating—and thence dealing with—these. By engaging with the artwork the audience is lifted—imaginatively and emotionally—out of the immediate continuum of personal existence into one informed by a more universal dimension of aeonic becoming. This is far enough from everydayness to be *release* but near enough to be *real*. It is our secular immortality and redemption'.[62]

What Crowther emphasizes here is the physically embodied artwork as a *made* object. In bringing the artwork into being, the artist makes individual choices and configurations of his *material*, which are contingent in the sense that other choices and strategies of configuration could have been chosen. In relation to the *finished* artwork, however, all these contingencies are rendered unnecessary. The identity of the finished work entails all those particular choices and configurations which were actually involved in its making. If any of them had not occurred, then a different work would have been the outcome. But it was not; we have just this work. There is no way in which we can know what the artist thought and felt in the process of creation, or the exact order in which the elements of the work were brought into being. But we do know that—given the finished product—there was a specific play of activity, a process of bringing forth whose success or failure is evident in the complete work.

If the work engages us, we can identify with the process of creation as a surge into being, a bringing forth which has been successfully accomplished, and which, in its success, serves as an exemplar of how the finished product of artifice can redeem and preserve individual moments of striving. This is a metaphysical aspect of what Crowther otherwise calls 'aesthetic empathy'. In it we identify with the process of artistic creation in *virtual* terms. It is this recreation of the virtual artistic process as opposed to mere fantasies of the actual process, which produced a specific work, which Cassirer needs in order to resolve the potential contradiction, which we earlier noted.

This emphasis on what is achieved in creating the work as an art object feeds back importantly to Chapter 2. There we considered how Crowther emphasizes the creative dimension of imagination and the narrative of the self. We insisted that the transience and frailty of the moments in our inner-life are things that we would like to make endure, and share with others, and, if possible, take to a higher level. It is precisely this that the making of art achieves. As Crowther shows, in the making of the artwork, the holistic structure of experience is embodied in a form that enables aesthetic empathy.

The other major problem facing Cassirer is another one that is also left over from his treatment of Enlightenment and Kantian aesthetics. It is the issue of the universal validity of aesthetic judgments. Earlier in the present chapter we gave consideration to the way in which Cassirer misidentifies the universal communicability of the aesthetic judgment with the fact of the artwork's objective intelligibility. This is a misidentification because the former pertains to the validity of our evaluations of aesthetic objects, whilst the latter pertains to our ability to recognize what a work is representing. This latter factor has now been clarified adequately by reference to Crowther's exploration of the concept of iterability and its significance in artistic contexts.

But given this, we surely do need something from Cassirer on what—if anything—justifies the critical judgments, which we make about the merit of artworks. A provision of this sort is even more pressing in the light of Cassirer's anxiousness to distinguish between art and mere entertainment. It is also demanded because of the unsatisfactory way in which Cassirer deals with this problem in the context of Kant's aesthetics. We will recall that he gets into difficulties by focussing too closely on the connections which Kant makes between art and genius. He would have been better advised to look at the objective expression of this relation, namely originality, as a definitive property of fine art.

Fortunately, at the end of *An Essay on Man,* Cassirer does address this possibility. He notes that in the context of art we are not satisfied with works, which merely repeat traditional forms; innovation is required. The point is made in even more detailed terms in *The Logic of the Cultural Sciences* (published two years before *An Essay on Man.*) In this very condensed work we are told that:

> No artist can represent nature without, in and through this representation, give expression to his own I; and no artistic expression of the

> I is possible without setting before us something objective, feeling and gestalt must merge into one another and must be completely absorbed in one another if a great work of art is to be created.[63]

Cassirer emphasizes that, in this respect, Goethe was right to emphasize the importance of artistic style. It means that art involves a productive attitude that must create new things rather than simply redeploying old material in a new guise. This is illustrated emphatically by the greatest works. They each leave us with the sense that we have made a new discovery in all sorts of ways.

Great poetics works, for example, do not merely evoke fleeting moods and impressions, rather they render those states through a linguistic form that exceed the scope of ordinary concepts and presents experience in a new yet enduring way. And in the visual arts. Michelangelo's *David* and Raphael's *School of Athens*[64] focus attention not on the physical materials of which they are made, or on their spatial arrangement, but on a broader representational function that allows the viewer to engage with the artist's own creative vision.

Cassirer also identifies the relation that is at the heart of originality in art. 'The creative process must always satisfy two different conditions: on the one side, it must tie itself to something permanent and existing, and on the other, it must always be open to a new employment and a new approach that changes this existence. For only in this way does one succeed in doing justice to the demands of both the object and the subject.'[65]

Cassirer continues,

> There is a vocabulary of forms of the painter, the sculptor, the architect, and there is a peculiar "syntax" for each of these fields, as there is a syntax for language. All this cannot be freely "invented". It is here that tradition repeatedly assets its claim; for only through it can that continuity of creative work, on which all intelligibility rests, even within artistic language, be established and ensured.[66]

No artist has mastered his or her medium if this mastery is not based on both the technical material conditions, and *form*. By 'form', Cassirer has in mind both tendencies inherent to the medium itself, as well as specific historical realizations of these. These are what is transmitted to successive generation of artists *as* tradition.[67]

However, we now reach the key area of difficulty for understanding art as symbolic form. As we have just seen Cassirer at last confronts the normative basis of artistic originality—centering on the individual artist inheriting forms that are basic to the tradition, and adapting these to his or her personal and historical circumstances. To put these points in the terms with which we have developed Cassirer's position, the existence of art depends on an interaction between the traditional forms of an iterable code, and the individual artist's innovative personal inflections of these. There is a key difference from the sciences here. Cassirer notes that:

> It is true that a great scientific discovery . . . bears the stamp of the individual mind of its author. In it we find not merely a new objective aspect of things but also an individual attitude of mind and even of personal style. But all this has only a psychological, not a systematic relevance. In the objective content of science these individual features are forgotten and effaced, for one of the principle aims of scientific thought is the elimination of all personal and anthropomorphic elements.[68]

However, in contrast to this, artistic originality is necessarily tied to the creative individual—the 'genius' in Kant's sense.

Here Cassirer moves in the right direction—from the objective interaction of tradition and individuality, to the subjective cast of mind, which this interaction embodies. But he does not give us the means whereby this link between tradition and originality can be thought-through in philosophically adequate terms.

Crowther's theory of art does so. In particular, it converges on the very issues that Cassirer raises (without exploring) in *The Logic of the Cultural Sciences* and at the end of *An Essay on Man*. Crowther identifies three features of artistic originality that distinguish it from originality in other areas of human endeavour. These three features are consequences of a single fact. He describes this as the 'immediacy' of the artwork, but 'sensible individuality' is perhaps, a better way of describing it. Crowther illustrates what is involved as follows:

> To enjoy art *qua* art, the starting-point for appreciation or enjoyment must be direct perceptual acquaintance with a particular artwork. This is because qua aesthetic object we are concerned with the formal relation between the parts and whole in the object's entire phenomenal fabric. Hence, if we are to enjoy the work's status as a formal configuration produced by artifice, we must focus on how the creator's sense of form structures its phenomena totality.[69]

From this three inferences can be made. First, that whilst the content of say, a philosophical or a scientific work, can be generally comprehended through paraphrase or summary, this Is not at all possible in the case of appreciating original art. If we have encountered it only through reproductions or (in literature) through summaries, we are not in a position to offer a definitive appraisal of its merit. This is because to make such an appraisal it to judge its cohesion as a sensible individual. But such cohesion yields itself fully only through *direct perceptual acquaintance*.

The second feature of artistic originality dwells on the fact that as a *created* sensible individual it has an 'internal relation' to the existence of the person or persons who created it. In principle, someone could bring a scientific or philosophical theory, or a technological invention into existence other than the person or persons who actually did so, But as a sensible individual the artwork, cannot. Of course, this is the point made by Cassirer (as noted above), but Crowther explains why this claim is justified. He says,

> Originality in art is a function in part of the formal relation between all the elements in a sensible or imaginatively intended particular. And the choices which inform the creation of such an order—even the smallest decision—are the expression of an entire unique personal history. To change the smallest detail of that history is to change the resulting work.[70]

If Kant's life in the process of writing the *Critique of Judgment* had taken a course other than the one that it actually did, the book's intellectual content might also have turned out to be different in terms of style, and organization. But the basic philosophical theses that the book proposes would not necessarily be affected by this different life-course. This is because the work is seeking to give articulation to *general* truths. The artwork, however, is a sensible individual and this means that—given a specific work—if the course of its creation had followed even a slightly different path, it would have issued in a different work. Crowther's theory here is very provocative and requires, perhaps, more substantiation, but it is at least viable in principle and fits in extremely well with Cassirer's more schematic position.

The third inference from the original artwork's status as a sensible individual has already been discussed in detail. It is the fact that as a *made* sensible individual—something that fuses sensible content with the articulation of an iterable code—the work has a basic sense which transcends the immediate circumstances of its production, and allows, thus, an aesthetic identification with that vision of the world which the artist's configuration of the code sets forth. People can describe to use in verbal or literal terms how they experience the world, but the artwork *shows* us this. It is, in Cassirer's parlance, an 'image of passion'. We can thus respond to it on our terms as well as the artists.

Cassirer's theory of art in *An Essay on Man* is highly schematic (even if backed-up by reference to *The Logic of the Cultural Sciences*), but as we have shown, Crowther runs with the points, which are central to Cassirer, and takes them further. Cassirer's position now appears to be a highly viable one. As well as broadening Cassirer's position the connections with Crowther do so in a way that provide a substantial rejoinder to those alternative viewpoints to which Cassirer gives a regrettably cursory treatment.

The form of this rejoinder can be described as an assimilation rather than outright rejection. We will recall, for example, that Cassirer dismisses Santayana's link between art and entertainment. However, if we emphasize the significance of originality in defining fine art, one can distinguish this kind of art from the formulaic and lowest common denominator representations produced by the entertainment industry. (Crowther, indeed, casts this in terms of the distinction between Art with a capital 'A' and art with a small 'a'). And again, whilst with Bergson and Nietzsche, we can agree that art might well sometimes have its origins in psychological states of a quasi-hypnotic, 'intuitional', or intoxicatory kind, what is distinctive of artistic creation as such is the kind of artefact which is produced rather than the kind of mental or emotional state which leads the artist to produce it. These states play an important role, but not the definitive one.

Matters lie similarly with the analogy between art and play. Art cannot merely be reduced to what is involved in this analogy but it does, at both subjective and objective levels, involve an element of play. In what we described as one of Crowther's 'finest' passages (and quoted at length earlier on) he presents us with the dialectic of contingent episodes in the making of an artwork. The painter, for example, does not *deduce* his or her next brushstroke or series of brushstrokes, there is an element of play and freedom in how they come to be juxtaposed with one another in the process of creation. It is only when the painting is complete that they assume the character of necessity.

But from the viewer's standpoint this necessity is not restrictive, rather the stable unity of the finished work allows free perceptual exploration to be guided gently, rather than to disappear in *chaos* of mere subjective associations. We can imagine in a free way the process of brushstroke on brushstroke, which allows the work to emerge into its finest forms, or we can identify with the style in terms of which the artist treats his subject matter. In all these factors there is an element of play exactly because we attend *to how* the work emerges, and *how* its forms cohere variously rather than simply judging *that* it is a painting, or *that* it represents such and such a subject matter.[71]

V The Historical Dimension of Aesthetic Cognition; Further Links to Crowther

In *An Essay on Man* it is notable that whilst Cassirer's offers an outline theory of the art object and our aesthetic response to it, he barely addresses the way in which art—as a symbolic form—is progressively articulated in historical terms. This is why Crowther's approach is so important. He does not simply define art as an aesthetic object produced by human artifice, but also as an *original* aesthetic object. This entails that the definition of art involves a comparative dimension, in that originality in a work is intelligible only by virtue of contrasts with other works. In practical terms this means that any judgment about whether something is *Art* or merely *art* must take account of tradition. Our horizon of comparison will be substantially of a *historical* kind. The historical aspect is integral to aesthetic cognition.

Because of his use of this comparative horizon, Crowther's definition of art is a normative one.[72] Some representations count as Art because of the ways in which they differ from other, more formulaic representations. They are significant refinements or innovations in relation to the tradition of making representations. It must be emphasized that their artistic/aesthetic power consists not in changing the medium's practices for the sake of it, but in order to create new ways for artist and audience to experience the world.

Now, Cassirer's link between symbolic form and its progressive articulation can itself develop Crowther's theory even further. The latter's aesthetic theory and the former's general philosophical strategy can thus be brought into alignment so as to achieve a relatively comprehensive theory of visual art as a symbolic form. We say 'relatively comprehensive' because of some weaknesses in Cassirer's position. The first is that (as already shown) he does not say very much in *An Essay on Man* on how we might understand the progressive articulation of art.

The second weakness is more wide-ranging. In fact, it is one of the most fundamental problems that Cassirer must face. The problem is indicated by the following remarks concerning the individual art media:

> 'Each of them proceeds differently; each sets a different accent of 'meaning' it extracts a different essential world of form. Painting, sculpture, architecture—they all go their own way, and the specific way of rendering space that prevails in each of them corresponds to a specific way of rendering objects. It is the task of systematic aesthetics to follow out all these particular ways of rendering and to show the characteristic 'principle' that they obey'[73]

Here, Cassirer is talking about distinctions between visual art media. However, whilst Cassirer points out the fact of differences of structural meaning both in particular terms and more generally, it is only at the most general level that he describes these differences in detail (for example, in distinguishing between, language, myth, and science, etc.).

But we need attention to the more particular differences as well. This is especially the case in relation to art. Art in its most general sense includes literature, music, dance, and the visual arts at the very least. However, if art is normative, then it will not be (as it were) a mini-symbolic form in the way that these media are, individually considered. Rather it will be one that is emergent from them. For example, literature will be language that is articulated in a symbolically distinctive way—neither reducible to mundane uses of language nor to idioms of meaning in the other art media. And within literature, there will be key symbolic differences between poetic meaning, storytelling, and theatre and the like.

Similar considerations hold in relation to the visual arts. There is a basic symbolic form—visual representation—with further differentiations such as picturing, sculpture, photography, digital imagery, conceptual work, and (at the intersection with functional significance) architecture. However, the visual *arts* emerge as a symbolic form on the basis of the style and originality (whose general terms were described earlier) in terms of which these media are articulated by creative artists.

It follows, therefore, that if we are to rejuvenate aesthetics in the vital directions that Cassirer's and Crowther's works point us towards, then, an important further step must be taken. We must identify those criteria that are basic to these individual media, and discuss the more detailed criteria of style and originality which allow them to become art. In particular, it is important to emphasize how the arts become meaningful through their relation to cognition, imagination and the aesthetic narrative of the self. In terms of this first task, Crowther's recent work[74] has taken up the challenge. He has identified the symbolic structures that give a distinctive character to all the aforementioned individual media. We will now consider this in relation to further possibilities opened up by Cassirer's ideas. This is an especially important part of our study, because it is here that a rewarding future for aesthetics can be glimpsed. A Cassirer-Crowther aesthetic cognition approach engages with the artwork as an artefact that transforms how the world appears, through the ontology of its medium, and style of its creation. The dynamics of artistic creation becomes the main object of attention. Through this, aesthetics becomes relevant again.

Notes

1 Ernst Cassirer, "Mythic, Aesthetic, and Theoretical Space", trans. Donald Phillip Verene and Lerke Holzwarth Foster, *Man and World* 2, no. 1 (1969): 3–17.
2 *EM*, p. 137.
3 *LCS*, p. 173.
4 Ibid., p. 139.
5 Johann W. von Goethe, *Goethe on Art*, ed. and trans. J Gage (Berkeley and Los Angeles: University of California Press, 1980), 109.
6 *EM*, p. 142.
7 Ibid., p. 143.
8 Ibid., (For some anticipations of this position see *PSF*, Vol. IV, pp. 78–79).
9 Ibid., p. 143.
10 Ibid., pp. 143–144.
11 Ibid., p. 144.
12 Ibid., pp. 144–145.
13 Crowther, *AE*, p. 19.
14 *EM*, pp. 145–146.
15 Ibid., p. 146.
16 Crowther, *AE*, p. 19.
17 *EM*, p. 145.
18 Ibid.
19 Ibid., p. 145.
20 Ibid., p. 147.
21 Ibid., p. 148.
22 Ibid., pp. 148–149.

23 Ibid., p. 149.
24 Ibid., p. 151.
25 Cassirer also considers this question in *PSF*, Vol. 4, p. 84 where he declares that '[f]rom the recognition that the human mind cannot attain the pure, clear realm of the visible form of things as long as it tries to grasp them only through immediate or reproduced perceptions we are lead to the additional insight that artistic activity is necessary to bring out this form and make it visible'.
26 *EM*, pp. 151–152.
27 See, for example, Immanuel Kant, *The Critique of Judgment*, trans. J.C. Meredith (Oxford: Clarendon Press, 1973). The main discussions are in Book I pp. 44–45, and pp. 65–67.
28 Crowther, *AE*, pp. 19–20.
29 For a fuller description of Cassirer's position here, see Chapter 3 of the present work.
30 *EM*, p. 152.
31 Ibid., p. 153.
32 Ibid., p. 154.
33 Susanne K. Langer, *Feeling and Form: A Theory of Art* (London: Routledge and Kegan-Paul, 1953), 48.
34 Ibid., p. 50.
35 Ibid., p. 72.
36 We are giving consideration to the edge of the plane here so as to accommodate the fact that even 'minimal' non-objective works can be read in relation to the notion of virtual space. This will be important later on in our book.
37 Langer, *Feeling and Form*, p. 109.
38 Specifically, see p. 154.
39 *EM*, p. 157.
40 Ibid.
41 *EM*, p. 159.
42 George Santayana, *The Sense of Beauty* (New York: Scribner and Sons, 1896), 22.
43 *EM*, p. 159.
44 Ibid., p. 160.
45 Henri Bergson, *Time and Freewill*, trans. F. Pogson (London: George Allen and Unwin, 1959), 14.
46 *EM*, p. 163.
47 See, for example, the *Introduction to Metaphysics*, trans. T. Hulme (London: Macmillan, 1913), 56.
48 See, for example Friedrich Nietzsche, "The Birth of Tragedy" included in *The Philosophy of Nietzsche* (New York: The Modern Library, 1954), 947–1088.
49 See *EM*, p. 164.
50 See, for example, Friedrich Schiller's *Letters on the Aesthetic Education of Mankind*, trans. E. Wilkinson and Willoughby (Oxford: Oxford University Press, 1984).
51 *EM*, p. 166.
52 It is significant that, since Cassirer's *An Essay on Man*, at least one other philosopher namely Hans-Georg Gadamer—has made an important connection between art and play. For Gadamer, however, play is given a rather deeper interpretation than the views which Cassirer is rejecting. It is seen as a clue to the ontological structure of the artwork, in that the 'truth' which emerges in the artwork is not a static content, but rather a dynamic one which arises through a complex play between world, subject matter, the artist's style and the tradition of creativity to which it both belongs and modifies. To explore the tension or possible harmony between Cassirer's and Gadamer's approaches would require an extended and difficult analysis extending beyond the purviews of this study.

For Gadamer's position see his *Truth and Method*, trans. William Glen-Doepel (London: Sheed and Ward, 1979). Of especial relevance are pp. 100–104.

53 *EM*, p. 169.
54 See, for example, *LCS*, pp. 116–120.
55 Crowther, *TLTCA*, p. 11.
56 Ibid., p. 20.
57 Crowther, *AE*, p. 20.
58 This does not stop us from appreciating an artwork in absolutely disinterested terms, but to do so does not do justice to its status as art.
59 Crowther, *AE*, p. 186.
60 Ibid., pp. 186–187.
61 See *EM*, p. 149.
62 Crowther, *AE*, p. 177, see also *TLTCA*, p. 12.
63 Cassirer, *LCS*, p. 31.
64 *LCS*, p. 43.
65 *LCS*, p. 116.
66 Ibid.
67 For Cassirer, the cultural historian Aby Warburg illuminates key aspects of this transmission in both historical and psychological terms. In *The Logic of the Cultural Sciences*, for example, we are told that 'Warburg originally set out from the history of art in the Italian Renaissance. But it constitutes only a single paradigm for him within which he sought to clarify the particular nature and general tendency of the creative process in the fine arts. He found both most clearly expressed in the survival of the image forms [*Bildformen*] of antiquity He showed how antiquity created specific pregnant forms of expression for certain typical, continually recurring situations. Certain inner excitations, certain tensions and solutions are not only captured in them, but they are, as it were, exorcised in them. Wherever a similar affect is discernible, the image, which art has created for it, comes to life again. According to Warburg's expression, certain "formulas of pathos" come into existence that are indelibly stamped into the memory of humanity. He explored the mutations, statics, and dynamics of this in detail. This has enriched the methodology as well as the content of art history.' The pathos-formulae which Cassirer alludes to here are specific stylistic motifs—associated, for example, with the rendering of drapery, that carry an affective charge that endures over the generations, and is central to the classical tradition in the visual arts. (*LCS*, p. 116).
68 *EM*, p. 228.
69 *AE*, p. 184.
70 Ibid., p. 185.
71 These recognitional factors are, of course, necessary conditions of our appreciation of a work, but they are not sufficient ones.
72 Crowther's most recent statement of it in these terms explicitly is his *DACC*, 2007. However, it is found throughout all Crowther's books, although he—rather misleadingly—describes it as an 'ecological' definition of art in his earlier works.
73 *PSF*, Vol. 4, p. 80.
74 See especially his *POVA*.

5 Visual Art and Symbolic Forms

Introduction

First, it is useful to recall the most basic idea of Cassirer's philosophy of symbolic form. It is the point that an individual sign or unit of consciousness has its identity only insofar as it functions within a more general rule-governed space of other such signs and units. The individual's identity is 'pregnant' with reference to the surrounding field that makes it intelligible.

The question arises as to whether we can trace this function more concretely in relation to the different media in which artworks are produced. Cassirer, as we have already seen, regarded each medium as having its own structure and syntax, but does not provide much detail of how this is so. In this chapter, we will explore the ways in which Crowther's philosophy of visual art allows this basic point to be understood in rich and illuminating terms, and we will show how its potential can be developed further by using our own examples.[1] Particular emphasis will be assigned to Crowther's ideas considering how spatial material is worked into different art forms, and is progressively articulated. This diversification is, *in itself*, an expression of progressive articulation, because as new media are explored, they extend the scope of artistic meaning.

A special strength of Crowther's approach is that, like Cassirer, he does not simply take symbolic forms as given, he goes back to their origins in basic features of embodied cognition. This approach, indeed, shows how art answers needs of experience that arise from the nature of subjectivity as outlined in Chapter 2.

I Architecture

Crowther emphasizes the extreme importance of the fact that an embodied subject is by definition an occupant of space. This very simple fact has great ramifications in many different directions. For example, the fact that the human being occupies space means that it has to engage in different ways of unifying its different aspects so as to survive in the face of the vicissitudes of life. Visual art offers ways of doing this, and does so on the basis of features unique to the individual medium.

For Crowther, the basic symbolic structure of architecture starts from the enclosing and/or declaring of space for functional purposes. This necessitates the connecting and/or working of three-dimensional material. Pictorial art is three-dimensional in mainly virtual terms but, of course, architectural works are physically three-dimensional. Sculpture is like this too, but architecture is three-dimensional in another way insofar as the architectural work can be physically occupied and moved through, as well as being perceived as three-dimensional. Occupancy and the possibility of shelter mean, of course, that architecture has a functional relevance that other art forms do not.

Crowther, however, distinguishes between building and architecture. As soon as someone lives in a community where there is choice as to how shelter can be constructed from available materials, then the situation has already gone beyond building per se. How the building is constructed becomes a significant consideration in its own right. Once the need for shelter or other modes of occupancy are satisfied then a variety of psycho-social needs require satisfaction also. This is the point—although Crowther does not himself declare it—when Cassirer's basic symbolic function comes into play. From this point building becomes more than building when it is locatable within a field of meanings that exceed building's merely practical functions.

It relates, for example, to choice in terms of such things as shape, materials, and location, and the way in which such essential features as roofing, walls, light sources, ventilation, and entrances, are distributed and articulated individually, and in concert. By means of choices concerning these, an element of aesthetic preference as well as practical concern enters into proceedings. The individual building is now pregnant with a space of displayed features which give it a different significance—as architecture rather than building.

Location is also central to architecture's distinctiveness. Through being constructed in a location, the work's structure and appearance have to engage not only with the physical characteristics of the location, but also with the visual appearance of nearby buildings and immediate environmental features. This constrains how an architect designs a work. If it is created without reference to its surroundings, it runs the risk of incongruity.

Crowther emphasizes that—in contrast to the major visual art forms of painting and sculpture—architecture is a non-autographic visual art form insofar as the architect does not have to be involved in the physical making of the edifice from his or her plans. However, whilst the work is non-autographic, it is nevertheless an individual, rather than a token of a type. This feature and the non-autographic character combine to give architecture a quite distinctive expressive character. It is connected to the nature of the individual architectural whole which involves material that has to stand erect, carry loads, and resist the forces of gravity and other physical demands. It must be given physical form in a way which not only fulfils its practical function but invests it with broader symbolic meanings.

Features such as regularity, symmetry, proportion, and harmony, for example, are key in terms of the disposition of part to part, and to the whole, in all organic structures and, most especially, the human body. In architecture, we look for similarities and contrasts in how such features are proportioned and arranged in relation to one another (in terms of both the building's function, and its appearance). We look, in other words, for individual organic unity on broad analogy with the human body itself—in terms of such things as posture, its ways of physically carrying burdens, and other modes of deportment and action.

For Crowther, then, the analogy with the body and its activities is central to architecture. He takes the analogy further still. A building's articulation of form, shape, and mass in enduring materials on the basis of symmetry, proportion and the like, in effect, idealizes the body's directions of sensorimotor activity. Sometimes, we lose our balance, or are overburdened, or are unable to make sense of something given in immediate perception. Architecture, however, involves the achievement of order and clarity in concrete and enduring terms. Giving form to massive and resistant material such as stone entails that a feature basic to gestural artifice—the giving of form to matter—is exemplified with especial drama.

Crowther relates these points to architecture's non-autographic character.

In architecture, the work is defined mainly through the creator's design, rather than (in contrast with painting and sculpture) the effect of his or her gestures. This has a special significance. For architecture is not the human body, nor is its physical fabric constructed through the architect's own gestures. Hence, whilst its forms have those significant correlations with the human body just noted, its expressive meaning will lack the gestural specificity that allows the close recognition of emotions. Describing a building as 'cheerful', 'melancholy', 'exuberant', or 'sad', is possible, in principle, but it would be strange to think of it as a basis for the general consideration of architecture's expressive dimension.

We are on much better ground when the viewer identifies with more general and very basic vital feelings such as elevation, illumination, mystery, security, belonging, or whatever. These feelings are not attributing of specific expressive states to the building itself, they are, rather, 'mood' responses that are closely linked to specific stylistic and structural features of the building, its mediation of light, and its opening and enclosure of specific spaces within the whole, as well as the effect of the whole, itself. In this way, the non-autographic nature of architecture allows it to operate at a more general and basic expressive level than autographic visual media.

Crowther argues, then, that architecture involves space created with the interests of bodily exploration as well as practical functions in mind. The deep affinity between the body and architectural space enables the subject to animate the building's forms, in distinctively expressive terms. Here the interplay of objective spatial presence and subjective experience is mediated by that aspect of the subject—the body—whose being is primarily spatial.

What is most important here is not just refinement or innovation in terms of the medium's structural principles, but the way this is achieved. When the artist refines or innovates in relation to one of art's symbolically distinctive media, the achievement is not a merely technical one, it is one that—literally—changes how the world appears, and in so doing, relates us to ourselves, the artist, other people, and the world in a transformed way.

Now, there is one striking feature that Crowther's theory of the arts does not develop, but which Cassirer's basic symbolic function points towards. As we have seen, that function is basic to knowledge of self and world, and is adapted and developed as circumstances demand, through time. This is why symbolic forms and their progressive articulation arise, phylogenetically. However, there must also be an ontogenetic aspect to this. The individual learns from circumstances, and can develop his or her self-understanding.

In the case of art, this process can surely expect to be even more emphatic, because art involves important dimensions of self-exploration achieved through working in a medium. Stylistic development is itself a mode of personal progressive articulation, which when it engages with important features of the relevant medium, also advances it.

In this chapter, then, as well as considering Crowther's approach to different art media we will take it further by consideration of how individual stylistic achievement can progressively articulate the relevant medium itself. We start with the architect Fillipo Brunelleschi.

His first architectural commission was the *Ospedale degli Innocenti* or Foundling Hospital that he worked on from 1419 to around 1445. Brunelleschi and Donatello had visited Rome at the beginning of the fifteenth-century, and the present building is perhaps testimony to the effects of that visit. Its façade is composed of an arcaded loggia, with the arcade itself constructed from round arches supported by mixed order columns (with roundels in the spandrels above them). The first and last bays of the arcade are framed by pilasters. At the top level are windows surrounded by pediments marking the centers of the arches.

The classically inspired simplicity of this design was a strategy that served Brunelleschi well as his career developed. *Sagrestia Vecchia* (the Old Sacristy) in Florence (worked on from 1421–1440) is a good example. Composed of four square spaces: the largest one is organized around the tomb of Giovanni di Bicci and his wife Piccarda. It is in the form of a perfect cube, above which is small hemispheric umbrella dome. Light enters through arched windows in the upper registers of the walls, through twelve round windows at the bottom of the cupola vault, and one other round window (high up). A smaller square space houses the altar, and two lateral small sacristies form closets.

An entablature runs all along the main interior space and the altar square, with 'support' from pilasters in the corner areas. The entablature and pilasters do not provide real physical support. Rather their function is aesthetic—to provide a system of rhythms that divides the space into two horizontal areas of equal measure. This, in turn, has the overall effect of emphasizing the geometric unity of the spatial layout of the building *in toto* by visually integrating the dome with the proportions of the square spaces below.

The aesthetics and technique of Dome construction are key features in Brunelleschi's greatest achievement—the Cathedral of *Santa Maria del Fiore* in Florence. The building had already been designed in the fourteenth-century, but the question of how to construct a Dome of the requisite scale presented a major aesthetic, and even greater technical problem. Existing techniques were not suited to the task, the more so in that the existing design ruled out familiar devices for supporting and articulating great masses (such as, for example, flying buttresses and pointed arches).

Brunelleschi's solution consisted of building two concentric shells, an inner one that would be visible to the congregation within the Cathedral, and a taller, external one, visible from outside. A further formidable problem—counteracting the bulging outward pressure—was solved by binding the walls with tension rings constructed from iron, stone, and wood (somewhat like the hoops on an old fashioned barrel). The lower regions of the Dome were made from stone, and the upper areas from lighter materials, but without using conventional ground-based scaffolding as a means in the construction process. (In fact, Brunelleschi invented and patented a hoisting machine for raising masonry to the upper reaches of the Dome.)

The double shell structure is light and slim both in visual appearance and physical arrangement, with an external 'skin' of terracotta cladding. It is created as an octagonal drum with a span of 140 feet, composed of eight segments (the 'sails') arranged on two shells with a space separating them. At the summit of the Dome is the 'lantern' some 375 feet above ground level

In all his work, Brunelleschi was inspired by Classical models, but his work is much more than a rediscovery of antiquity. We have already seen how he combines elegantly rational visual composition with technical innovation. Through this, he changes the nature of architecture itself. He progressively articulates the medium. However, the important issue—as in all cases of progressive articulation, is how these developments enhance our aesthetic experience of the world and the self.

In consideration of this we will recall the act-association continuum—that real of frail and transient imagery in which our everyday practical consciousness is immersed. All architecture offers an aesthetic compensation for this transience. In Brunelleschi's case, the Dome of *Santa Maria del Fiore* is a physically massive edifice which, at the same time, radiates the simplicity of its formal organization. Human ideas and inspiration may be frail and passing, but they can be expressed through or settle on, perceptually speaking, that which is more enduring. In Renaissance architecture, indeed, the stability of geometric forms not only offers a compensating perceptual

Figure 5.1 Filippo Brunelleschi, *Cupola of St. Santa Maria del Fiore*, Florence, 1436. Image by courtesy of Sailko

solidity, but also, paradoxically enough, often evokes a sense of the fluidity of movement amongst them.

Brunelleschi's interiors are particularly effective in these terms, using pilasters and false entablatures to both link and demarcate basic 'real' physically embodied geometric forms such as squares and semi-circles. In this way, the cold geometric structure of spatial mass is animated—is made to echo the life of spirit within the material body. The geometric forms are experienced simultaneously as things to be perceived, and as spaces which can be occupied. Indeed, whilst Brunelleschi's work declares the reality of geometry as an abstract mode of symbolic form—where steps of proof emerge discursively; these are, nevertheless embodied to the senses as aesthetic transitions, living structures that energize the real. Here Cassirer's basic symbolic function is expressed simultaneously under two different aspects.

II Pictorial Art

By creating pictures, humans present features of both the spatial world and how they interpret it. Most importantly, they do so at the level of space itself. This means that pictures allow personal interpretation to be made publically available. However, in contrast to architecture, and sculpture, the third dimension of pictorial space is virtual rather than real.

Crowther emphasizes that there is a close connection between mental imagery and the pictorial image. As we saw in Chapter 2, mental images play an important role in our knowledge of the self and the objective world, but, from the viewpoint of psychological introspection, they are flickering, fragmented and stylized. Things are changed when the ability to make images is exercised in a physical medium, and the outcome is a picture. The resulting image is stable and unified. Even more important, the artist's style is taken—through handling, and composition—to a level where it is more developed, and, through culminating in the creation of a physical object, is brought to completion.[2]

In terms of what we called the act-association continuum of experience, pictures have a special significance on the basis of Crowther's approach. Our experience is never able to stand still in moments of self-containment. However, pictures at least give the illusion of doing this. It is as though *the artist has eternalized a moment of perception*, and transformed it into something with enhanced meaning. It is a symbolic answer to the frailty and transience of experience, as such.

Crowther explores many aspects of the uniqueness of pictorial art. He argues that the special relation between subjective image-forming, and pictorial representation, is a major reason why pictures have intrinsic fascination, and great communicative power. At the most basic level, a picture is the projection of a three-dimensional item or state of affairs within a virtual two-dimensional plane. The referential basis whereby we can see the marked surface as a three-dimensional something is some element of visual resemblance in terms of shape, texture and/or color, between the picture and that which it represents.

Once this natural phenomenal relation of visual resemblance is codified as pictorial representation, there are potentially an infinite number of pictures which can be made on the basis of it. However, as Crowther shows in the greatest detail, there are only a finite number of logically distinct ways in terms of which pictorial space can be composed. These are the categories which define the general scope of picturing's symbolic articulation of the image.

However, whilst Crowther takes these categories to be logically implied in the basic ontology of picturing, he emphasizes that their practical emergence involves very complicated interactions between different pictorial media, diachronic history, and the individual and historical circumstances of the artists who articulate them. Mathematical perspective, for example, is first hinted at in classical antiquity, but develops only as a major compositional criterion in the tradition that extends from Giotto to Masaccio. That perspective emerged under these circumstances with these particular artists is contingent. Indeed, there was no necessity that perspective should ever emerge at all. The fact that it did so is down to contingent historical circumstances; the fact that it could emerge is down to the ontology of pictorial representation.

Once formulated, mathematical perspective dominated how pictorial space was composed in the European tradition for the next 450 years or so. It has a flexibility and comprehensiveness which allows that which is represented to be situated in relation to a virtual infinite system of spatial coordinates. There is no connection between this and how we see in strictly optical terms, although Crowther identifies an important relation to spatial perception as such. It arises from the fact that if an artist wishes to wishes to represent three-dimensional states of affairs in a two-dimensional-plane so as to be as consistent with the systematic connectedness of physical reality itself, as it is possible to be (in a static monocular image), then mathematic perspective is the means to this. It is more than a culturally relative construct. Perspective offers a mode of unifying pictorial space that is most consistent with presenting a visual analogue to the way that the spatial world reconfigures systematically in correlation with movement through it.

Giotto, the Lorenzetti brothers, and many others, abandon the principle of planar unity that formed the basis of medieval art. Their paintings create historical transformations that culminate in Masaccio's achievement of mathematical perspective. This achievement brings about a qualitative and paradigmatic change in how pictorial space is explored. It dominates until the end of the nineteenth-century when art rediscovers and transforms a more planar conception of pictorial unity, This, in turn, allows Picasso and Braque and twentieth-century abstraction, to take pictorial space into an entirely new area.

What we have just been describing is not confined to pictorial space alone. Pictorial art has many other structural features that are taken to higher forms through historical development. These concern such things as modes of delineating—e.g., with a painterly or with a linear emphasis, the way forms are organized in relation to the edge of the picture, and a host of other things that we have explored in detail elsewhere. However, whereas, for example, in scientific theories of space, new theories come along and replace others at the cutting edge of understanding, in the arts, pictorial styles and tendencies that are no longer fashionable, i.e., always retain a powerful aesthetic potential despite having gone into the past. Individual artists can find ways of renewing or transforming features that have already emerged in pictorial art, and giving them new stylistic expressions. (We will return in more detail to this very important point in the Conclusion to our book.)

Crowther's position, then, sees pictorial art having a basic structure where the individual work gets its meaning through its position in relation to the diachronic history of pictorial media as well as to referring on the basis of selective visual resemblance. In this it perfectly exemplifies Cassirer's basic symbolic function. Progressive articulation is achieved when creative individual works engage with and realize key structural possibilities inherent in the medium. By doing this, they allow the visual world, and the artist's 'feeling' of it, to be expressed in original terms.

We will consider the example of Raphael here. In his *Marriage of the Virgin* of 1504, Raphael adapts a compositional structure from his (probable) master Perugino. It has traditional elements—such as the frieze-like distribution of figures in the main frontal plane. Perugino's treatment of the subject arranges the figures in a somewhat stiff way, whereas Raphael's figures are linked through an elegant shallow structure of intersecting diagonals that gives the group a vitality that suits the joyous subject matter. Here we see a significant advance in stylistic terms. Perugino's basic compositional format has artistic value, but Raphael allows its potential to developed even more

In other works before his move to Rome in 1508, Raphael takes his independent approach further still albeit in a way that continues to negotiate other influences, notably that of Leonardo. However, in the Stanza della Segnatura frescoes painted for the papal apartments of Julius II, Raphael finds both an environment of work, and compositional vision that leads to his greatest achievements. Notable amongst these is *The School of Athens*, of 1509–1510.[3] The fresco is adapted to the physical demands of a large apartment, but the virtual space it projects is that of an exhilaratingly lofty basilica—a built space whose meaning simultaneously alludes to Roman civic life, and papally sanctioned spectacular ecclesiastical buildings. The narrative is a complicated one that centers on Plato's pointing towards the heavens, and Aristotle towards the earth. This need not be read as a conflict because philosophy has to make sense of reality in terms of both its otherworldly metaphysical grounds, and its existence as we negotiate it through the life of the body.

One of the strongest features of the fresco is the way Raphael connects the figures in specific conversational groups, yet, at the same time, also conveys a sense of these groups also being linked in the greater conversation of philosophy itself. The unity here is both formal and iconographical. It

Figure 5.2 Raphael, *Wedding of the Virgin*, 1504 (Pinacoteca di Brera, Milan)

centers on Raphael's wonderfully calibrated combinations of color and gesture in the figures, the positioning of the groups, and the relation between the airy light and elegant stages of planar recession in the transition from the foreground space to the background. Raphael based many of the figures in

Figure 5.3 Raphael, *School of Athens*, 1511 (Stanza della Segnatura, Vatican City)

the work on his own acquaintances, and one's overall sense is that Raphael is not simply creating and populating a virtual space, but is projecting a cultural milieu in which he feels personally at ease, an ease which translates into the felicitous compositional and color strategies that allow the viewer to feel aesthetically accommodated also.

This great strength in the combination of color and design is the basis of Raphael's originality extends the scope of the medium. All painters have to deal with the relation between design and color. Raphael achieves a balance between the two of such grace and vitality, that it becomes exemplary—a standard for other artists to assimilate and to work from. For centuries afterwards, in academic teaching, Raphael's mature works serve as a standard of excellence for the color/design relation.

The phenomena of progressive articulation in the individual artist's style can also be vividly emphasized through the example of Picasso. His early Blue and Rose period works (1901–1904, and 1904–1906, respectively) have a sparse and plaintive approach to color, composition, and narrative elements. There are some loose precedents for this in work by Puvis de Chavannes, but even in this early period Picasso's work is distinctive, His emphasis on angularity and schematic approach to faces, along with his low key treatment of light, color, and texture, makes his realistic or 'matter of fact' kinds of subject matter disturbing.

If Picasso had never developed his style further, these early periods would have been achievement enough. However, in *Les Demoiselles d'Avignon* of 1907, he makes one of the most radical innovations in the history of western art. Here, what was an angular emphasis in some of his earlier works, is now made into a space-constituting power, with both the female forms and those of their physical setting created from facets, with minimal tonal effects that actually suggest the striations on carved material. The consistency of conventional (virtual) pictorial space is disrupted by lPicasso's formal and compositional strategy.

The Cubist works by Picasso that follow *Les Demoiselles* reduce three-dimensional things and states of affairs to basic geometrical solids, whose volume is expressed through broad facets rather than color and texture. By 1911, as well as three-dimensional forms being reduced, they are also formed from broken facets in a way that allows them to merge with the surrounding features. Indeed, in a work such as *Ma Jolie: Woman with Guitar or Zither*, the virtual three-dimensional substance of the sitter and her environs are totally broken up and redistributed in a system of trellises and triangular structures. This is abstract art. However, Picasso always insisted that his art engaged with the real, and the inclusion of lettering at the bottom of the work, does re-establish the link with familiar visual reality, although it is somewhat tenuous.

Even without the lettering, and considered as an abstract work, *Ma Jolie*, and similar paintings, are very instructive. On the one hand, the particular aspect represents not only hidden aspects of the represented objects / states of affairs, but also the artist's act of appropriating and transforming the scene; on the other hand, the particular configuration suggest, by association, the kind of visual details and textures that are present in all visual things but which we do usually notice. In terms of this second feature, all recognizable objects are recognized as emergent from a space of hidden aspects, or suggestions of relations to other things or values. The given object points towards this context, and gets part of its meaning from it, although we might not be explicitly aware of all the factors involved. According to Crowther, it is precisely this *transperceptual* space that allows us to find abstract and neo-abstract art such as Picasso and Braque's hermetic Cubism both meaningful and important. (Indeed, further on, we will return to the importance of the transperceptual in the context of literature.)

There is one further achievement in Picasso's artistic development that we must consider. In 1912, he took another momentous step with his *Still Life with Chair Caning*. Braque had already been experimenting with creating images from the juxtaposition of drawings with other physical materials. Picasso took this into the realm of painting by laying down an area of oil cloth (with a chair cane design) onto a painted canvas. This radically changes the whole enterprise of visual art, by allowing works to include visual material that was not created directly by the artist. (As we shall see later, Duchamp takes this much further.)

The point of importance to take from these considerations is the way that Picasso's individual—sometimes almost playful—self-expression is done in a way that impacts on the scope of the artistic medium itself. He takes pictorial art to the verge of abstraction, and then changes it again, in a way that (as we shall see) makes Conceptual art possible. The example of Picasso again shows that progressive articulation within an artist's style, can, if that style has radical implications, change the scope of the medium in question. The medium is advanced.

III Photography

Crowther argues that photography has its own unique symbolizing structure. Photographs work by causally capturing a single visible aspect of their subject. Autographic media such as drawing and painting create their subject matter through the artist's physical gestures. He or she creates shapes, lines, colors, and textures, and the like, to produce an image of some kind of thing. Sometimes, we may know that the picture is based on an actual existing individual three-dimensional object or state of affairs (or one that formerly existed), but this knowledge has no necessary bearing on recognizing what kind of thing the picture represents. The question of whether the subject actually existed or not, is secondary.

With photography, the reverse is the case. That which the photograph represents is—through its creation from mechanical and chemical processes—a direct causal trace of the referent's visible being. The photo may be faked, but if it is a photograph at all, then, by definition, it must be created from the physical traces of something that actually existed at some time or other. Crowther explains, in other words, that photography is a *causally rigid* mode of visual representation. Photographic images present the causal traces of actually existent or formerly existent space-occupying phenomena. The relation between what a specific photograph shows, and our knowledge of the causal rigidity of photography as a physical and symbolic idiom, is the basic symbolic function which makes it one of the symbolic forms of visual art.

Embedded in this relation are other important symbolic factors. We often think that that the changing spatial and temporal positions of something's existence are contingent matters, individually considered. Had things been different, the existence in question might have involved very different spatial and temporal positions. However, according to Crowther, once a being has moved into the future, its past states—no matter how small-scale they may have been—are necessary factors in determining the character of the events that come afterwards. If we want to explain how the being exists in the future, then we must include all its past states.

It follows, accordingly, that the photograph's causally rigid relation to its subject matter not only shows that that subject once existed, but also declares some of the passing visual aspects that played a necessary role in shaping its future identity and appearances. Through being featured in the photograph, visual aspects that otherwise would be unnoticed are preserved from disappearance into the past. The photograph not only, in effect, asserts that this subject matter once existed, it also discloses features which figured, necessarily, in how that subject matter's movement through space and time developed. By separating these moments from the subject, it allows them to achieve an, at least, symbolic autonomy that our actual lived-moments of experience cannot achieve individually. This (and the compositional criteria we will mention in a moment) mean that photography is a mode of spatial formation rather than a passive mechanical reproduction of the spatially given.

Photography is a symbolic form, then by virtue of our reading of the photograph not as a picture, but as a causally extracted scene from a spatiotemporal continuum of real life events, of which, the taking of the photograph was one. The photograph exemplifies a unique form of the basic symbolic function by virtue of the relation between the individual print and this continuum of the real.

This basic function can be progressively articulated in artistic terms because whilst photography is causally rigid, it also depends on choice in terms of what is photographed and how it is photographed. The artist can select spatial subject matter that appears strictly perspectival or which focusses on landscapes (or whatever) with spatial recession that is ambiguous and which might even suggest aggregate space. He or she can compose in a way that selects subject matter that appears to recede in gradual planes or which cuts through the planar structure with more dramatic effect. Of the greatest importance, also, are techniques for adjusting focus and for creating and developing the negative.

The career of Alfred Stieglitz is especially useful in showing aspects of the progressive articulation of photography (although Crowther does not discuss him). By the late 1880s Steiglitz had already acquired a deep knowledge of the technical and chemical aspects of photography, and was advocating its potential as an artistic medium. Acquiring a hand-held camera in 1892, offered him the opportunity to explore an expanded area of visual possibilities.

In the 1890s and first decade of the twentieth-century, Stieglitz operates in what is often described as a *pictorialist* idiom. Put simply this is an approach that seeks to create photographs by 'tweaking' both reality and the techniques of taking and developing the image. By selecting a view that looks as though 'it ought be in a picture' and by using the appropriate—in effect, painterly—technical means, the photograph is presented as if it were a work of pictorial art. It gains aesthetic legitimacy, in other words, by looking towards a different art medium. In *The Terminal* of 1893, for example, Stieglitz's image involves a horse-drawn New York tram, with the tram itself and the horses present along different oblique axes. This compositional device is overlain by steam arising from the hot bodies of the horses in the cold air. This misty effect gives the work a quietly urban picturesque effect. It is a photograph that wishes to become a painting.

Stieglitz created many such—often superlative—images, perfecting the approach compositionally and technically through experiments in drawing, waxing, and toning on platinum prints. However, in *The Steerage* of 1907, he advances much further (although he himself did not apparently realize the importance of this image until the following decade). The photographic content is structurally organized around the presentation of an inclined funnel, and a gangway bridge that projects upwards and obliquely from the middle left of the photo to the crowded upper deck. Below it is an equally crowded lower deck. The visual details of the individual figures and their distribution is such as to create a kind of rhythm of visual push towards, and pull away from the viewer, that engages our attention more than the literal spatial relationships of the people in the photograph and their physical setting *qua* physical

Figure 5.4 Alfred Stieglitz, *The Steerage*, 1907. Image by courtesy of the Google Art Project

The importance of this is how it creates a composition of shapes and ambiguities of spatial relation without any 'dressing up'. The image has no obviously poetic potential, that would give it pictorialist meaning, and the camera and print make no use of special effects. The image is rooted in three-dimensional visual space and recognizable occupants of it, but it selects a particular visual perspective with a richness that unties the ordinary vectors that organize our normal vision.

This reveals a truth. Photography is grounded in the real—it has the causal rigidity that Crowther highlights, with all the intimations of mortality that are consequent on that. However, Stieglitz's image shows that photography can be an artistic medium without resorting to the model of painting. Our perception is allowed to shift from formal to physical aspects to virtual content, without any arty prompts. *The Steerage* derives its aesthetic power from the easy movement back and forth it allows between a causally rigid capture from the real visual world, and the freedom in how we explore this visually. The artist discovers a reality that he or she does not create, but having found it, allows it to release its aesthetic potential simply through the right kind of positioning. This is the artistic feature unique to photography. The importance of Steiglitz's photograph is that it is one of the earliest works to realize this potential in an insistent and accessible way. It is, accordingly, a major step in the progressive articulation of his own style, which also has an equal importance for the artistic advance of photography itself.

IV Digital Imagery as Symbolic Form

We turn now to Crowther's interpretation of digital art. Digital art with a visual orientation is based on electronic technology that makes use of mathematical algorithms for acquiring, storing, processing, and displaying visual information. This sort of information can be displayed in either three-dimensional or two-dimensional formats. The digital 'image' takes the form of material generated electronically from mathematical formulae (including those used to convert already existing images into a format that is decidable by the right kind of hardware. Digital imagery can involve both images and words.

Such imagery can be generated at as many different times and places as there are devices to perform such generation. Whilst, like photography, digital works are tokens of types, their imagery is different from photography in a key ontological way. In photography, there is always at least some difference between each token *qua* physical object, however slight. The type which the digital work instantiates, in contrast, is of the same ontological order as its tokens—it is a programme constituted from mathematical relations and algorithms. *Qua* programme, it is created as a type before its tokens, but in every other respect, is identical with them.

This relation embodies Cassirer's basic symbolic function. The programme is one that can be realized an indefinite number of times in exactly the same form. Digital imagery has two basic formats of realization. It can be done *discretely* on individual computer screens or through using larger, computer-directed, Liquid Crystal Display (LCD) formats. The other major form of realization is through what Crowther calls *distributed* modes of digitally generated imagery or visual environments. These do not have to be on a single screen. They can, indeed, use computer directed-optical effects such as Light-Emitting Diode (LED) lights, or be generated through the linking of multiple screens and optical devices.

Again, it must be emphasized that what makes digital imagery worthy of progressive articulation are further symbolic features embedded in the relation between the programme and its idioms of realization. Some digital modes create artefacts with an autographic aspect, or which appear to have been painted. Sometimes this latter effect is produced through the computer-directed placing of actual paint upon a surface. Hybrid works like these can define forms with astonishing clarity and sharpness, and blend them with extraordinary smoothness, and visual complexity.

These works radicalize how the represented content appears. Crowther argues that this radicalization is the key to a symbolic structure that is unique to digital imaging—namely visual *morphing*. Its significance can be understood first by contrast. 'Identikit' images are familiar things—being composite images made from one image placed on top of another. The morphed digital still-image, in contrast, is a composite structure(s) generated through a single, *continuous* electrical/mathematical process. The end product is in that way freed from the appearance of having been created out of disparate source elements. Scanned material can be used, but the emergent image is more than the sum of the original scans that it was derived from. It has what Crowther describes as 'holistic visual structure'. What is key, here, is not skill in how the image is realized but ingenuity in the original programme.

The holistic distinctiveness of digital morphing is expressed most completely through temporally realized modes. They involve visual transformations in linear successive time, but not in the same way that video and film achieve such effects. The latter can involve cross-fading and the like only through strict limitations. These are based on the fact that temporal realization in film and video involve the successive presentation of *individual* photographic frames at an appropriate speed. The individual frames are causally related to an original visual state of affairs—even if the use of special effects and editing try to disguise this relation.

Digital imagery, in contrast, is generated without any causally rigid relation to what it represents. Such a relation is, at best, involved indirectly when the imagery is based on scanned material. Even here, however, the relation is changed. Once an image is digitized and incorporated in a programme, it becomes, in principle, manipulable and realizable in ways that differ from the appearance of the sampled or scanned original. As we already noted, film and video can be transformed by editing or addition, but digital works contains their diversity within a single programme.

Crowther's point, then, is that temporally realized digital morphing is qualitatively distinct from the principles of animation that are basic to film, video, and, of course, cartoons. However fantastic the traditional media may be, the elements from which they are generated are constrained by the real, in a way that digital morphing is not.

There is a second important feature that Crowther takes to be distinctive to digital works. It is their *interactive* dimension. There are precedents. Some surreal or conceptual works are premised on response and cooperation from their audience. Such works, however, are the exception rather than the rule. With digital works, in contrast, the interactive dimension not only is more frequent, but also can involve an original programme that is modifiable, in principle, wherever and whenever the work is electronically realized.

Again, for Crowther, this involves a qualitative change in terms of interactivity. He identifies five main vectors of interactivity.

i) local and non-local interface—the former involving a user in physical proximity to the computer, the latter, a user who achieves remote access
ii) where boundaries of interaction are precisely defined by the artist. Here, user choice is confined to pressing 'Enter', opening or shrinking windows, choosing between links, or other simple operations. No significant feedback of a creative kind is required from the user. He or she simply uses the device to open up possibilities created in advance by the artist.
iii) interfacing based on computer responses to an audience, occurring whether or not the audience is aware that an interaction is actually taking place (e.g., when the computer generates imagery in response to movements by the audience, or to other environmental variables)
iv) voluntary interface based on the user's *navigation* of a programme. Here the user exercises choices which the programme on the computer responds to by opening up new creative possibilities for the user to respond to in turn.
v) user-transcendent interface—allowing the user to initiate and guide a programme in a way that allows it to develop autonomously—without significant further input from the designer or user

These distinctive realms of morphing and interactivity mean that digital imagery involves a vital new form of visual art with its own new principles of creativity. The primacy of programme-design involves issues that far exceed mere mathematical or technical competence. For example, the morphing dimension has to visualized and considered in terms of its structural and developmental scope—i.e., in terms of the character of its visual realization(s). In the case of interactivity, its navigational forms allow the viewer/user of a programme to explore and extend its scope through choice and—in the case of games—an element of skill.

When it comes to finding principles of progressive articulation for digital imagery, matters get much more complicated than with the other visual art forms just considered. The interactive dimension and constantly developing technology may involve key principles of their own.

Crowther himself points out features which may transform the visual arts. The traditional work, for example, is based on both individual creators and physically individual products. Revisions or additions to the work are not allowed—except under very special circumstances, and, even then, only if carried out by the original creator. Computer generated works in contrast, start from a defined individual programme but one which is open to evolutionary change and the participation of many individuals or even groups, other than the originating artist. The emergence of navigable and evolutionary digital works, in other words, opens new possibilities of collective authorship and open-ended works whose realization is never brought to completion.

In terms of insights concerning the progressive articulation of digital art, we are at a very early stage of development. An artist such as Charles Csuri, for example, first used computer assistance to create drawings and paintings in the mid 1960s, but by 1996 had assimilated techniques for generating images exclusively from programmes. In effect, these are algorithmic paintings. In them the seamless transition involved in temporally realized digital art's morphing finds a counterpart in the hyper-real effect of the static image. There have, of course, been super-realist tendencies in painting as well since the 1960s, but the precision of the algorithmic image is such that its electronic origin is inescapable.

One artist is of special interest in illuminating the relation between individual progressive articulation, and that of the medium itself. He is William Latham. His work uses programmes (or genetic algorithms) that mimic natural processes.

The origins of this approach are found in Latham's student days when, during studies at the Natural History Museum he devised drawings of evolutionary processes and considered the possibility of combining different animal forms using basic visual 'operators' such as 'beak, bulge, twist, scoop, grow tendril, and stack', as evolutionary principles to organize the appearance of new visual life forms.

On the basis of the aforementioned drawing, after computer training he and Stephen Todd created 'Mutator' software. This allowed the computer-generated appearance of possible evolving life forms based on the artist's selection of forms using genetic algorithms. In this, Latham expanded his original operators to include shapes and forms derived from combinations of plant and animal forms, and related natural phenomena. These were combined using Darwinian principles for a kind of natural aesthetic selection. The operators now involved concepts such as good, bad, marry, breed, kill. Crowther notes that

> As a starting point for digital creation, Latham would choose an aesthetically attractive three-dimensional form, and then randomly mutate its 'genes' (i.e., the software code sustaining the form) so as to create (usually) nine variants. One of these would then be selected for further mutation, and the process repeated again and again until some culminating 'whole form' that struck the artist as being an aesthetically fitting outcome to this process of quasi-natural aesthetic selection.[4]

Latham's *Mutator* series cross boundaries between distinctions between the creativity of humanity, and of nature, and of computing. They issue in stunning images of possible life forms that, as with much algorithmic images, declares itself as hyper-real and electronically generated. The special individuality of Latham's imagery is described by Crowther as follows:

Figure 5.5 William Latham, *Mutation X-Ray92*, 1992. Image by courtesy of the artist

> its insistently real appearance is at odds with the fantastic and unlikely character of its configuration. In the tension between these, the idea of alternative organic form is opened up for imagination. Technology here naturalizes itself through the projection of alternative possible life-worlds, on the basis of quasi-natural techniques of evolutionary generation.[5]

Latham has continued this evolutionary aspect of his work into the field of temporally realized and *interactive* digital work. This means that the progressive articulation of his own style is one that has emphasized the two factors that are basic to the progressive articulation of digital art itself as a unique medium. Indeed, his approach does not use computing as a means to create hyper-real algorithmic 'paintings' but to evolve images. He uses creative principles whose combinations and development are only possible through computing.

We have shown, then, that Crowther's approaches to pictorial art, photography, and digital art are exemplary in expanding the scope of Cassirer's account of symbolic form—although they were devised over sixty years after his death. We have also gone beyond Crowther in showing how progressive articulation within an individual artist's oeuvre can also lead to progressive articulation of the medium itself.

Crowther does not cite Cassirer in his recent work. The theories we have just described, however, not only share the same general philosophical agenda, but also *meet in spirit*.

This being said, there is an interesting objection that might be raised. Whilst Crowther's work allows Cassirer's philosophy of symbolic form to be extended to the visual art forms individually, it surely cannot encompass one of the most influential art idioms of all—namely conceptual art? Actually it can—and Crowther again shows us how.

V The Unexpected Example of Conceptual Art

Conceptual art sometimes consists of texts alone. For Crowther, this makes it rather self-defeating as specifically visual art. If conceptual work is to be genuinely visual, how it looks has to be an integral part of its meaning. Its character as *this* visible thing specifically cannot be dismissed as a mere contingency. Otherwise there are no logical grounds for thinking of it as visual art at all.

Much more interesting material is provided by the tradition arising from Duchamp's ready-mades where spatial items or states of affairs made by persons other than the artist can be appropriated for his or her artistic purposes or incorporated into his or her work. This basic symbolic act—of simply designating some object or state of affairs to be art, occurs within, and thence (in Cassirean terms) is 'pregnant' with, a field of broader meanings which give it intelligibility as an individual artwork. Crowther suggests that found material can be used to create 'conceptual phenomena' with a special form of visual interest that has its own aesthetic worth. More specifically, Crowther holds:

> 'The phenomenon is realized when the artist actually produces the visually interesting dimension through modifying or conjoining found material. The fact that the found material is still manifestly an aspect of what is thus made, means that the artefact is different from art as traditionally conceived—it has an 'idea' aspect. But, at the same time, it is configured in such a way that its visual distinctiveness as a particular, or token of a type, is essential to understanding its 'idea' in the fullest terms'.

Surprisingly, Crowther provides a set of categories whereby the basic symbolic function formulated by Duchamp is progressively articulated. In what follows we offer a slightly revised version of this, and a number of examples that Crowther does not consider.[6] The categories are:

i) unassisted found objects—where something is designated 'art' on the basis of being displayed on the basis of an art display format or in a context where 'art' expectations are The work requires no further physical interventions on the artist's part. This, is the basic symbolic strategy originated by Duchamp and manifest in a work such as *Bottlerack* of 1914.
ii) assisted found objects—where an artist or his or her agent changes the found object to greater or lesser degrees without totally concealing its original identity. An early example of this is *Bicycle Wheel* of 1913 which was, chronologically speaking, the first ready-made. However, it counts as an assisted ready made because Duchamp did not simply display it, but mounted it on a specially adapted kitchen stool. One of the interesting ways in which Duchamp articulated the ready-mades was in raising questions about what counts as 'assisting' them. The famous *Fountain* of 1917, for example, is a porcelain urinal, but is assisted to become an object of aesthetic attention through being signed and dated 'R. Mutt 1917'.
iii) manufactured objects—where the artist provides instructions for someone else to make an object which the artist then designates to be his or her own artwork and not that of the person who made it. In 1935, Duchamp designed the *Rotoreliefs*—double sided discs that, when spun on a turntable at 40–60 rpm, give rise to a depth-illusion. The Rotoreliefs were manufactured to Duchamp's specifications, in an edition of 500. This strategy—of designs that were physically created by others without any physical input from the artist—has been of great importance for minimal sculpture, as well as for future conceptual works.

iv) fixed assemblages—where found or manufactured objects are physically changed and connected permanently to other objects to form a complex single work, or a serial one composed from more complex individual units. Duchamp's *Why Not Sneeze, Rose Srelavy*? of 1921 is an example of this. This takes the form of a birdcage that contains a thermometer, a cuttlefish, and 152 marble cubes. The cubes were commissioned by Duchamp, and shaped by others.

v) moveable assemblages—complex works which are created by being assembled from found objects or other material but which can also be disassembled and installed elsewhere. For example, Duchamp's most important later work *Etant Donnes* (which he worked on between 1946 and 1966) involves a mannequin of a naked woman lying on her back holding a gas lamp, in an urban parkland. This scenario can only be viewed through a peephole in a wooden door in front of the tableaux. The work is made from an arrangement of found materials, and constructed features. Whilst *Etant Donnes* was probably intended by Duchamp to be exhibited at one site, it can, nevertheless, be disassembled and reconstructed, in principle, anywhere.

vi) site specificity—when a work created in any of the foregoing ways is created for installation or allocation to a particular physical site or location for a specific time period only. In 1942, for example, Breton staged a retrospective exhibition of Surrealist art in New York entitled *First Papers of Surrealism*. Duchamp used string to thread a giant web through the entire exhibition space. This artificial structure was adapted to the exhibition space—a former mansion house which included gilded mouldings, ornate ceiling paintings, and crystal chandeliers, as well as the paintings. Whilst it is possible to repeat this web-creating procedure in any space, as an act, it is tied as an inaugural strategy to this time and space in New York.

vii) found bodily activity—actions performed for some other purpose than those associated with traditional gestural art idioms such as dance or mime. The performance may or may not be recorded—as the artist wishes. In some cases, recording the performance is regarded as an essential part of it. Duchamp's transvestite persona 'Rose Srelavy' is an example of this. 'She' features in a number of photographs taken by Man Ray in the 1920s, and Duchamp himself used the pseudonym to sign works or written notes. The name itself is variously thought to be an anagram of '*Eros, c'est la vie*' or '*arroser la vie*'. Other artists also, such as the poet Robert Desnos, used the name as a basis for playing linguistic games exploring the relation between meaning and nonsense. More generally, Duchamp's assumption of the Srelavy persona—in his act of dressing for the photos, and posing as a woman, displaces transvestism from the domain of sexuality and /or theatre, whilst retaining elements of

> both. In particular Duchamp's gestures as Rose allow the persona to become an informal institution for creative acts that explore the boundaries of meaning, and what those boundaries will tolerate.

Crowther takes formats of these kinds to describe the basic conceptual possibilities available for creating Designated Art from Found materials. However, we emphasize much broader implications. For example, the very first of these possibilities—where art is created by an act of designation rather than an act of making—has a logical priority, although the creation of assisted ready-mades preceded it. It acknowledges, indeed, that the basic symbolic function is not only the basis of meaning; it is also the basis of creativity. This is because Duchamp here takes something familiar and by linking it to an unexpected context invests it with a new and imaginative meaning. As soon as this gesture has been made, all the other possibilities of conceptual art are opened up as explorations of the possibilities opened by this creative act. The chronological order does not matter here. What matters is the way the act of art-creation by designation is isolated and then applied across a range of contexts.

The various principles involved in this are, in Cassirean terms, the progressive articulations of Duchamp's unexpected way of deploying the basic symbolic function. It is not a simple linear development, but rather a fund of ideas— ideas that have become strategies adopted by other artists in the twentieth-century and beyond. Duchamp's inaugural gesture could have simply be taken as anti-art. But that it developed a tradition in its own right is testimony to it being symbolically pregnant with other creative possibilities—of creating spatial formations by using or arranging individual objects that need not themselves have been physically made by the artist. Duchamp's own stylistic development as an artist itself maps out the basic directions that progressively articulate a new visual medium that opens up creative possibilities for others.

V Conclusion—The Varieties of Visual Symbolic Form

We have described how Crowther identifies basic symbolic structures that are distinctive to the spatial being of visual art. He clarifies the formative spatial power of the visual arts to a degree that other thinkers (including Cassirer) have not done. By doing this, he allows all the major visual arts to be integrated in a lucid and insightful way within Cassirer's philosophy of symbolic forms. His treatment of conceptual art and related tendencies is especially important in respect of this integration. Because Cassirer's own theory of art emphasizes its sensuous character, it might be thought that the conceptual art realm must completely elude it. Crowther's account, however, actually assimilates its most important aspects within the terms of symbolic form. Where we differ from Crowther, is in our emphasis on how the development of an artist's individual style—his or her creative achievements—are the features that also bring about progressive articulation of the medium itself. The ontogeny of an individual artist's stylistic development can—when it engages with fundamental features of a medium—bring about phylogenetic change that advances the medium.

It will be recalled that in Chapters 2 and 4, we emphasized the psychological need to overcome the transience and frailty of mental states and images by embodying them in an enduring medium. We have shown how Crowther's approach reveals that the visual arts do this in a privileged way by creating space-occupying images whose spatiality is a part of their meaning. More than this, they articulate the image further by adapting it to the demands of the medium *qua* symbolic form, and by transforming this, in turn, on the basis of individual style.

This leads to a final vital transition. We have shown how different visual media are individual varieties of the basic symbolic function's spatial expression. However, we will also remember that Cassirer regards art *as such* to be a symbolic form. But art encompasses literature, theatre, and music (amongst other media) as well as the visual art forms. And one might expect that each of these non-visual art forms will have their own internal varieties which again express the basic symbolic function in distinctive individual ways. There are, in other words, symbolic forms based on the ontological distinctiveness of the media involved, then there are higher symbolic forms which share properties at a group level (e.g., the visual arts), and then, more general still, there are symbolic forms based on relations between individual examples that cross group and ontological boundaries. Art *as such* is the main example of this.

However, let us now consider major examples of the arts of temporal realization following the Cassirer-Crowther approach.

Notes

1 In this chapter we follow the accounts of visual art that Crowther presents in the relevant chapters of *POTVA*. These are Chapter 9 for architecture; Chapters 2 and 3 for pictorial art; Chapter 8 for photography; Chapter 9 for digital art; and Chapter 7 for Conceptualism. The key discussion of the categories that define the logical scope of pictorial art, are proposed by Crowther in Chapter 4 of *TTI*.

2 Crowther's theory considering the relation between mental imagery and picture-making has been given most consideration in his paper "How Images Create Us: Imagination and the Unity of Self-Consciousness", *Journal of Consciousness Studies* 20, no. 11–12 (2013): 101–123.

3 Cassirer himself saw Raphael's picture as an exemplar of how art integrates the objective, the subjective, and the artist's individuality—"we can see in Raphael's *The School of Athens* nothing else but a canvas covered with flecks of color of determined quality and spatial arrangement. In this moment the work of art has become a thing among things, and knowledge of it is subject to the same conditions that hold for any other spatiotemporal existence. [However] The colors in Raphael's painting have a "representational function" insofar as they point

to something objective. We do not lose ourselves in their contemplation, we do not see them *as* colors; rather, we see through them something objective, a determined scene, a conversation between two philosophers. But even this objective content is not the unique and true subject matter of the painting. The painting is not merely the representation of a historical scene, a conversation between Plato and Aristotle. For in truth it is not Plato and Aristotle who speak to us here, but Raphael himself. These three dimensions—the physical experience, the objectively represented, and the personal expression—are determining and necessary', *LCS*, p. 43.

4 Paul Crowther and Isabel Wunsche, *Meaning in Abstract Art: Between Nature and Theory*, ed. Paul Crowther and Isabel Wunsche (London and New York: Routledge, 2012), 222. It should be emphasized that Crowther himself does not explicitly link Latham to the progressive articulation of the digital art idiom. For a highly rewarding discussion of the strengths and possible weaknesses of Crowther's analysis of digital imagery see John Lechte's *Genealogy and Ontology of the Western Image and its Digital Future* (London and New York: Routledge, 2012) (especially pp. 143–147). Lechte suggests that Crowther identifies the aesthetic aspect of digital imagery with those features which are ontologically unique to it, e.g., morphing. This is not wholly accurate. Ontologically unique features are clearly a necessary condition of the digital work's artistic status in Crowther's terms, but his definition of art is a normative one. To become artistic, the distinctive features must themselves be used creatively—that is to say, in an original way. Such a usage means that the image is used to explore experience rather than the politically loaded informational role that Lechte identifies. Interestingly, the origins of Crowther's normative definition are partly found in a source to whom Lechte is especially attentive—Walter Benjamin. In Chapter 4 of *CAP*, for example, Crowther makes extended critical use of Benjamin's famous paper 'The Author As Producer' for normative ends.

5 Crowther and Wunsche, *Meaning in Abstract Art*, pp. 223–224.

6 Crowther himself argues (*TPMA*, p. 180) that Duchamp's unassisted ready-mades are not aesthetically significant. This is a moot point. However, even if they are not they at least create the possibility of a new variety of artistic creation—one, indeed, which Crowther's other principles articulate.

6 Literature, Theatre, and Music as Symbolic Forms[1]

Introduction

At the start of the last chapter, we noted how Crowther does not simply take the visual artistic symbolic forms as given; he looks for their origins in more basic acts of meaning-giving. This makes his approaches to literature and music especially striking. For him, the origins of literature are in fine speech and writing. The point is that language can be used in ways that sets it apart from its ordinary descriptive and expressive uses. Sometimes these uses are tied to unique events, other uses are recurrent—to mark significant occasions. In either case, language is used—in terms of enunciation and content—in a more formal way than is usually the case for purposes of celebration or commemoration.

Traditionally, language used in this way is associated with rhetoric. The basic modes of rhetoric center on the use of vivid and well-articulated speech or writing for persuasive purposes. The intention behind this is encourage the acceptance of particular beliefs, or the taking of specific actions. Rhetoric is has especially close associations with religious, civic, and political discourse.

Rhetoric—like celebratory language—can be very much tied to the circumstances of its employment, and afterwards be of interest only from a historical perspective. However, sometimes rhetoric exceeds its original context or occasion of performance rhetoric exceeds its occasion or context of employment—for example, John Donne's sermons, Lincoln's Gettysburg address, or Churchill's wartime speeches). The writer's style of language in articulating the concerns of the moment or occasion, assume a more enduring expressive and moral meaning. There is a reciprocity between the particular speech' relation to the circumstances of articulation, and the fact that circumstances like those—such as the question of faith, or the vicissitudes of war, are recurrent problems in the human condition. In Cassirean terms the basic symbolic function is here engaged through the particularities of a particular speech-act, operating in a way that speaks beyond its original circumstances of articulation, so as to illuminate a more universal dimension of meaning that is at issue in them.

In order for rhetoric to engage the listener or reader's interest, it makes use of characteristic linguistic devices such as metaphors and metonyms. For Crowther, these devices are of special importance for they have an intrinsic significance that pushes language beyond its immediate indicative or persuasive goals towards characterizing that which they refer to. Metonyms for example, work by representing their object through reference to one or more one of its familiar properties, or with broader effects that are familiarly associated with it. These devices can be mundane and obvious— as when Paris is signified by reference to the Eiffel Tower, or in proverbial wisdom such 'too many cooks spoil the broth'.

However, there are metonyms which are more challenging because they are surprising and/or provocative—for example, the term 'screw-haired Thracians' in Greek poetry—where the phrase picks out a supposed feature of Thracians but also uses it to suggest that they are coarse or ugly.

In some cases, it is not the individual metonym, but their ways of being linked which sets up the imaginatively interesting meaning(s). An extract from one of Churchill's famous wartime speeches shows this strategy at work—

> We shall go on to the end, we shall fight in France,
> we shall fight on the seas and oceans,
> we shall fight with growing confidence and growing strength in the air,
> we shall defend our Island, whatever the cost may be,
> we shall fight on the beaches,
> we shall fight on the landing grounds,
> we shall fight in the fields and in the streets,
> we shall fight in the hills;
> we shall never surrender.[2]

A single powerful metonym can exert an immediate attention-attracting effect. Lincolns use of the phrase 'four score years ago and ten' at the start of the Gettysburg address is a metonym that uses the mathematical components of ninety (and an old fashioned term for describing one of them) to create a kind of linguistic fanfare wherein the passage of chronological time is poeticized as a setting for issues concerning historical destiny. The particular metonym or linkages between metonyms, in other words are imaginatively striking through their illumination of a wider context of cultural expectations and concerns.

The effect of metaphor is often especially powerful and has been very widely discussed in all traditions of aesthetics since the Second World War. Crowther notes with approval Paul Ricoeur's useful basic definition of metaphor in the following terms. It involves 'giving an unaccustomed name to some other thing, which thereby is not being given its proper name.'[3] It is sometimes also said that metaphor can also involve assigning a term to a subject that applies to it so obviously as to seem ridiculous to even point it out, for example, 'no man is an island'. However, there is a case for arguing that in cases of this sort, what we are dealing with is not metaphor at all, but actually a form of metonym. This is important, because it is clearly qualitatively different from the basis of metaphor—where a subject not only fails to receive its 'proper name' but is linked to a term that is manifestly incongruous or even incompatible with it. More specifically—and despite appearing as a deliberate error—the secondary subject works, in the most formal terms, as something asserted or 'predicated' of the principal subject. This idiom of predication, however, is an exceptional one. Nelson Goodman observes that:

> Briefly, a metaphor is an affair between a predicate with a past and an object that yields while protesting. In routine projection, habit applies a label to a case not already decided. Arbitrary application of a newly coined label is equally unobstructed by prior decision. But metaphorical application of a label to an object defies an explicit or unconscious prior denial of that label to that object. Where there is metaphor, there is conflict.[4]

On these terms, then, in metaphor, the predicated label introduces a formal element of incongruity. It is manifestly not the kind of thing which can be truthfully asserted of the primary subject. Expectations based on the conventions of language are at once and obviously affirmed by the predicative form of metaphor, but denied by its relation to our normal expectations. The basis of metaphor is creative tension.

Crowther offers a further explanation of metaphor. Because it transgresses conventional literal usage, metaphor has a conceptual connection with novelty in both objective and subjective senses of the term. The objective sense is the fact that at a metaphor defines itself as innovatory by standing out in relation to standard predicative practices surrounding its principal subject; the subjective sense is that innovation of this sort requires wit or inventiveness in the one who first formulates the metaphor.

This latter sense allows authentic metaphor to be separated from merely nonsensical sayings such as 'he's as angry as a bladder'. If this is to be more than nonsense, then there must be some logical connection between the secondary and principal subject. It is this connection that is, according to Crowther, the most fundamental positive feature of metaphor.

The connection is provided by a space of association. Crowther explains it as follows. The perceptual field, by definition, consists of things that are accessible to perception through movements of the body or changes of perception. Any individual perception or object of perception gets its individual feature through the relation between its own physical properties and position in relation to other things and their properties (including light levels and atmospheric conditions), and, just as importantly, to our anticipations of what would be revealed if we changed our position, or how things might have appeared had we occupied a different position.

Other factors that influence our present perceptual orientation include the relation between what is presently perceived and our entire personal history, and the cultural meaning of what we perceive or think about. All in all, things are not simply present to perception; they are recognized and situated in terms of different patterns of interest determined by all the factors just described. In Chapter 2, we will recall that it was considerations such as these that Crowther uses to argue for the unity of self-consciousness being understood in narrative and aesthetic terms. His treatment of metaphor emphasizes a particular aspect of this consideration. For Crowther, the perceptual field is subtended by the latent *transperceptual space* of unnoticed and or hidden phenomenal items and relations and a range of imaginative and cultural associations (a space that, as we saw in the previous chapter, is also—in its visual aspects—the ground of meaning in abstract art).[5] Everything that we perceive, think about, or imagine is given its character not only by individual features, but also by its relation to an infinitely complex network of contextual factors.

As far as we can tell, Crowther's choice of the term 'transperceptual' is to emphasize how nothing given in perception or consciousness is ever present as a self-contained atom.[6] To cognize it is to place it in a broader context, and it is only this placing that defines its individual character. The perceptual field is always being crossed by us, even when we are physically stationary.

Crowther's notion of the transperceptual also (although he himself does not remark on it) cashes out Cassirer's basic symbolic function as a complex cognitive operator that puts the individual item of perception in a *reciprocal relation* with the transperceptual domain. It is reciprocal in that just as transperceptual space informs how we experience the immediate object of consciousness, so that experience at the same time modifies—however slightly—the nature of the whole in which it is embedded.

The transperceptual dimension is worthy of a book length study in its own right because it takes us to the very heart of the uniqueness and complexity of our embodied subjectivity. It is also of the supreme importance for making aesthetics 'relevant' again, precisely because the arts engage with and clarify aspects of this very dimension.

In day-to-day perception and cognition, transperceptual space functions as a condition of cognition; it is not generally noticed in its own right. However, there are points in our use of language where it does begin to make its presence felt in more explicit terms. Metaphor is the main example of this. In metaphor, the subject of a phrase or sentence is *cognitively elucidated* by what is predicated of it, through the incongruity of the predication. Crowther holds that the predication 'works' because it directs towards transperceptual relations that occupy the associational space between the subject and its metaphor-creating predicate.

Consider, for example, Homer's metaphor of the 'rosy fingered dawn'. 'Dawn' is the subject although it occurs before the predicate 'rosy fingered', Dawn is a time of day, of course, rather than a body with hands and digits.

However, the inclusion of the term 'rosy' in the predicate terms is the catalyst for 'fingered' to work as a metaphor. The dawn, for example, can arise through horizontal bands of clouds that might suggest fingers; fingers are also used to create states of affairs, and in this sense there is a suggestion of the clouds bringing forth the day—and the beauty of the restoration of light each day; the mention of fingers in the context of dawn might also connote the intervention of the Gods, and the day being brought forth for a special purpose on this occasion. These features are all element in a space of transperceptual association stretched out between the terms dawn and 'rosy fingered'.

To recognize the relation between the predicate and subject term as metaphorical is to discern an analogy. In living metaphors we are challenged to imaginatively traverse elements in the transperceptual associational field which connect the subject and its analogical predicate. Through this, the principal subject is elucidated by virtue of our deepened experience of it, that is, the outcome of situating it in the field of transperceptual possibilities that surround it.

How one imaginatively traverses this field will vary from person to person. It is for this reason that metaphors cannot be paraphrased in literal terms. To be a metaphor is to posit an objectively significant connection, but how the associational space between the subject term and predicate is activated will involve key elements of personal association. In many cases, of course, a metaphor will be striking without provoking any occurrent state of imaginative association. But the very fact that we find it striking means that we recognize that it could deliver such associational richness if we were to dwell on it. To put it in Cassirean terms, we recognize its 'symbolic pregnance'. Whereas such pregnance is the basis of meaning *per se*, it is usually something that we do not notice in its own right. In linguistic devices such as metaphor, in contrast, its working is more overt in that attention is drawn to how meaning is created in the particular case.

The point is, then, that devices such as metonym and metaphor help invest speech and/or writing with an intrinsic interest. They can be used in rhetorical strategies, but also make language interesting in its own right, as a source of imaginative association. Metaphorical meaning is more than indicative. It invites us to consider *the character* of what is indicated through an imaginative traversal of the associational space between subject-term and predicate term, that draws on one's own experience of the sense of these terms.

The intrinsic pleasure of such meaning is one of the key origins of literature. Certain forms of language push towards the transcending of their indicative factual content or contexts of use because they use that provoke imaginative association which embeds them in personal experience. Of course, they are not forms of literature in their own right, and on most occasions we will just use them in practical communicative contexts. This is especially the case with dead metaphors, when the subject has been linked to its incongruous predicate so often, that the incongruity is no longer noticed (for example, in a metaphor such as 'falling in love').

We shall focus, however, on the contexts wherein devices such as metonym and metaphor are deployed within broader linguistic structures that stimulate imaginative deliberation and the enjoyment of how particular linguistic expressions create their meaning.

Of special importance for these structures is the relation to the narrative/aesthetic dimension of self-consciousness. We have emphasized the way in all experience involves narrative as a principle of unity that plays through imaginings and all those related features of the act-association continuum. We have also emphasized the frailty and transience of the states of consciousness involved in all this. It is important to declare now that even narrative unity itself is also elusive. We may have a broad sense of what we are that shifts and gets new inflections as our personal history develops, but we can never—through introspection alone—do justice to the many layered aspects of our narrative. However, a Cassirer-Crowther approach, clarifies how the arts intervene upon this. As we have seen, pictorial art and sculpture takes the image to a more complete state of realization. We shall now see how the arts of temporal realization—literature, theatre, and music—offer a similar realization of narrative meaning.

I Poetry

First, as we have already seen, the rhythms and cadences of language are expressive. Language is loaded with them not only to communicate information but also to reveal our own attitudes to what is being communicated (through such things as angry, joyous, sad, or surprised tones). These features involve variations of volume in how we speak, and also differences in rhythm. Rhythm in language involves measured exposition based on regularly distributed pauses, and patterns of repetition. It is these features that are basic to formal structures such as verbal rhythm, assonance, alliteration, consonance, and rhyme.

In ordinary circumstances we do not speak or write in such formally measured terms. But if we organize language in a way that does emphasize such features, then this acts as a kind of linguistic framing—declaring our utterance or writing as something that should be taken to have special importance. In celebratory verses—of the kind found in greetings cards, the formal structure is used evoke feelings. Indeed, the origins of poetry itself may be in ceremonial rite where language has to be elevated and made special so as to evoke communal feeling. Such cases, of course, use poetic language functionally rather than call attention to the means whereby the evocative effects are achieved.

However, poetry can also be developed in its own right, as a mode of language that invites consideration of the way in which this particular piece of language achieves its effects through individuality of articulation. Indeed, the origins of poetry and its major pre-Modern manifestations are in the use of formal structure and devices such as metonym and metaphor. This formalization can be used as a vehicle for more general ideas—as in the philosophy of Parmenides and Heraclitus, but this is only a use. The key to poetry is its employment of formalized structure as a characterization of its factual content from a personal point of view. This allows the poet's own conception of the worth and significance of the narrative content to be expressed.

Of course, it may be that the poet has disregard for the content—in which case he or she is not just disdaining it but emphasizing and expressing his or her attitude of disdain. Sometimes, indeed, a poet might also evince some degree of self-contempt (as is, perhaps, the case in some of Baudelaire's works). But paradoxically, by putting the negativity in poetic form, the poet is tacitly affirming that there is something of human interest in the comprehension of this very self-contempt.

The point is, then, that in poetry, the content and the poet's perspective upon it are made to inhabit one another in a way that solicits attention from the audience. Through this three-way reciprocity, the expression and experience of language is made special.

As an example of the individual poet whose stylistic development also progresses the development of the poetic medium itself, we shall consider Stephane Mallarme. Mallarme's work inherits the mantle of Baudelaire—that is to say a highly subjective vision of life that takes as its premise an *ennui* that arises not only through the failure of pleasures to satiate the sensibility, but also from a sense of the futility of life itself. In Mallarme's early poems, the sense of ennui is coupled to a frustrated yearning for transcendence towards absolute metaphysical meaning. In *Les Fenetres*, for example, a hospital patient gazing through a windowpane wishes to experience the sunset beyond the window directly

Et de m'enuir, avec mes deux alles sans plume
- Au risqué de tomber pendant l'eternitie?'
[and to escape with my two wings without feathers
- at risk of falling throughout eternity.[7]

In many of his poems, Mallarme, finds evocations of the metaphysical ultimate amongst aesthetically intense experiences. In *Prose; for des Esseintes*, for example, he describes iris flowers as follows

Telles, immenses, que chacune
Ordinairement se para
D'un lucide contour, lacune,
Qui des jardins la separa.

Gloire du long desir, Idees
Tout en moi s'exultant de voir
La famille des iridees
Surgir a ce nouveau devoir.

[Such, immense, that every one
was ordinarily garlanded
with a lucid contour, a lacuna
that separated it from the gardens.

Glory of long desires, Ideas,
Everything in me exalted to see
The family of irises arise to this new duty[8]

Here the opening word in the first quoted stanza—'Such' refers to the palpable being of the flowers, and is then elaborated in a way that emphasizes the subtle dialectic of their material being and ontological identity as a kind. The irises have a material fullness, that at the same time defines them individually—gives them precise contours. But just as this very precision separates them from the 'gardens' and makes them stand out, it does so by introducing a dimension of absence or negation. In effect, Mallarme is here poetically negotiating the Spinozistic/Hegelian notion that limitation involves negation. However, rather than dwelling on the metaphysical ramifications of this, Mallarme emphasizes the fact that in this confrontation between material fullness and limitation, a reconciling term emerges—the Ideality of the iris genus. As instances of kinds, all finite things have a unique essence which, in privileged moments is declared by the beautifully configured example. It is through this, that in the present case, Mallarme's irises rise, accordingly, to a 'new duty'. They are a mean of aesthetic/metaphysical revelation.

This revelation culminates in Mallarme's most celebrated individual work—*Un coup de des*. In the midst of the many iconographical complexities of this poem, a modest symbolic fusion of poetry and the absolute is achieved, they become a single fulfilling narrative. A clue to this main thematic is declared in capitalized text which is not readable as a single continuous phrase, but which is picked out through *looking at* the visual components of the phrase, whose culminating features are distributed across the poem. The full message of the work is then recognizable, visually, as 'A THROW OF THE DICE WILL NOT ABOLISH CHANCE'.

A throw of the dice is what any poet does, metaphorically, when writing a poem. He or she uses familiar words and phrases but must use these to articulate personal experience. But even when the poem is written, there is no guarantee that it will be successful—either in completely articulating what the poem was inspired by, or in communicating it to others. The poem is launched on a voyage which can end—to use a metaphor that is used in the Mallarme's poem itself—in *shipwreck*.

However, the abolition of chance for Mallarme is itself an aspiration—the one which we have charted briefly in his earlier poems. He wishes for some ultimate state of communion with the absolute—where all is fulfilment and necessity. Of course, this is not fulfillable in the life of a finite being, neither can it be fulfilled in poetry—although the poem can at least offer a kind of symbolic fulfilment of this. Mallarme's way of achieving this is of the greatest importance for poetry itself. He makes the 'look' of the poem—literally, how the words are physically distributed across the page—into a vital part of the poem's meaning. This is a vital achievement. The Other which the poet confronts—the physicality of the world and its alien character is drawn into the meaning of the poem itself. By being drawn in, the element of chance—elusive finite Otherness—is captured, and made reciprocal with the creation of the poem. Just as the poem emerges from the poet's engagement with the physical world, so the physicality of that world is here declared through the integration of space-occupying text and poetic meaning. The materiality of words on a page becomes an aspect of poetic spirituality. Of course, this is not a real abolition of chance, but, in Mallarme's terms, it is surely as near as a finite being can get, and it is achieved through poetry.

The most important aspect of the trajectory we have just described is not *Un coup de des* as a culmination of Mallarme's stylistic development in itself, but the way *this trajectory progressively articulates the nature of the poetic medium as such*. In order to deal with his metaphysical yearnings, Mallarme has to move the poetic vocabulary into entirely new directions—'paint not the object but the effect it produces' as one of his most memorable metaphors has it. In effect, this shifts the burden of meaning in poetry heavily into the Crowtherian 'transperceptual'. But whereas in the normal poetic use of metaphor, we imaginatively traverse the transperceptual associational space between subject term and predicate, in Mallarme's case, the emphasis falls on imagining how the subject term and predicate come to be linked at all.

However, there is even more to the transformations that Mallarme creates. In arriving at his final work, he also takes poetry into a further—and even more radical—new realm, where the physicality of the text is made poetically functional. This turns out to be an element of recurrent importance in twentieth-century modernist poetry. It changes the nature of the whole poetic enterprise.

II Storytelling

Storytelling centers on narrative. Narratives involve descriptions of how a certain train of events concerning *sentient beings* transpired. It would make no sense to tell 'stories' about, say, rocks and stones, unless they were invested with quasi-sentient qualities or which issued in outcomes that had a bearing on the lives of sentient beings. Characteristically, narrative will have a beginning, a middle, and an end, because this is the basic structure of all sustained actions—in terms of both physical gestures and an embodied subject's practical projects.

So as to distinguish between exclusively factual description and narrative, there is a further consideration. Narrative presents action as described by a witness, or *as if* it had been witnessed by the narrator. This can be literally intended—when the narrator tells the listener or reader that the account is based on personal experience, or is based on what he or she has been told about such experience. But just as often the narrative events can be described from the viewpoint of a possible observer 'as if' this was how they would have been perceived to happen. Narrative in either of these senses is the basis of storytelling. The narrative projects circumstances wherein the narrator's account engages his or her audience as though they too were observers of some chain of events that happened or could have happened.

In cultures without writing, it will be sometimes hard to separate storytelling from factual narrative descriptions. As we saw in Chapter 2, memory does not reproduce events and persons with absolute exactitude. Any recounting of past events—even recent one recorded in writing—will tend to involve some elaborations that go beyond the facts, even if this is not deliberately intended. Any fact can be described from different descriptive viewpoints, and sometime fictional elements will enter in. This does not have to be negative in terms of the truth. A truthful fact can be embedded in a tissue of fictive elaborations, and yet the upshot can still have a general truthfulness. *Many* recollections of events from our own past have this character. Indeed, as we also saw earlier, the basic features of subjectivity have an intrinsic aesthetic character as a result of the nature of imagination, and its relation to the narrative unity of the self.

This aside, there are some narratives that are overtly intended to be interpreted in 'as if' terms—as representations of possible happenstances, rather than records of something that actually happened. The narrator presents the story as a way of experiencing such and such a series of related events. The account may be based on actual fact or historical existence, but equally so, it may invent the content. Some of it may be so at odds with the way things normally happen that the fictional nature of the work becomes overt fantasy. (The criterion of 'normally happen' here, is in some respects, culturally specific.)

We must ask the question, then, of how it is possible to separate an overtly fictional narrative from one which is factual? In the case of fantasy, the criterion has already been provided. But how do we separate, say a vivid factual narrative from one that is mainly—and is intended to be—a work of fiction? One way is by analysis of the form and content of the text itself. Fantastic elements (such as supernatural events, or alien abductions) will count in favour of fiction, but are by no means conclusive, because many people have written such narratives in the sincere belief that they are true. This being said, there are certain ways of writing—especially recounting the detailed experiences and subjective responses of persons other than the narrator, which could not be based on what it is possible for a narrator to know. No one can experience how other people think and feel from the inside, so when a narrator talks on the basis of such knowledge, we can reasonably surmise that what we are reading is a work of fiction.

Of course, our sense that the work is a story rather than a factual narrative can be established by the simple fact that it adopts a presentational format which declares itself as a story or novel. But this does not have to be conclusive. We can conceive of the discovery of a narrative that is factual, but where there is nothing about the format of the text, or context of its discovery to show this. What is—for example—a vivid diary—might be discovered in a context which leads it to be taken as a work of fiction.

These worries aside, it is clear that there is a central core of writing where the idiom of narrative, and its mode of presentation, that establish the work as a fictional narrative. (The marginal cases mentioned above are *just* marginal.) *A story narrative is intended to engage our capacity for imagining. A description may do this, but it is not the basis of it as a communicative act.*

Some shared characteristics of poetry and storytelling should be noted. There may be some poems and stories which are told once without any intention that they should be repeated. But if something is expressed in language, it has the capacity, in principle, for being repeated on another occasion. Poetry and storytelling as practices, in effect, formalize this principle of repetition. This feature is enhanced, of course, when the story or poem is formulated in an exact, and stable order of words. It is because of this that literature has a different ontology from most visual art media. Paintings, drawings, sculptures, and the like are individual works. One can make copies of them, but in such a case the copies really *are* copies and not the original work itself.

Literature is different. A poem or story may exist in an original manuscript. The original can be recreated in different physical embodiments (as, for example, in the individual texts included in the print-run of a written work). Here, the differences between the physical formats in which the work is embodied are not logically relevant to the meaning of the work itself. Even in works such as Mallarme's *Un coup de des.*, whilst the physical organization of the printed page becomes part of the meaning, this does not extend to the physical differences that distinguish one copy of the poem from another. The literary work is a *type*—it has an exact number of specific words in an exact order of arrangement—of which its particular instances (or 'copies') are the *tokens*. The work *qua* type cannot exist unless it has at least one token, but there is no limit on the number of tokens in which it is physically embodied.

In order for the literary work to exist in the fullest sense it must be read and, thence, *imaginatively realized*. Through this, the work is read not only as a description but *as if* one were present at the events described, or, in some cases, as if we were the subjects of the experiences being described. The only constraint on this imaginative realization is consistency with the work's narrative.

Of course, each time a work of literature is read there can major imaginative differences between the present realization and other ones—a fact which holds even when the same person reads the same work on different occasions. Someone might say 'but each time *I* read a poem or novel, I always imagine what happens there in exactly the same way', but this claim is impossible to verify, and is at odds with the creative and unstable nature of mental imagery itself. What is important is the fact that significant variation *can* occur in different imaginative realizations of the literary text. This means that when we read a literary work, it becomes uniquely our own, even within the constraints of the guiding theme or narrative itself.

As an example, of individual achievement in storytelling that has significant ramifications for the medium as a whole, we consider Miguel Cervantes. His earliest novel *Galatea* is a pastoral tale that follows an already established idiom of following a group of people who, travelling, meet other characters along the way, each with a story to tell. *The Exemplary Novels* of 1613 is a collection of novellas that combine tales of amorous pursuit and entanglement with rather contrived plots, and very basic characterization. As in much fictional literature of the time, the main orientation of the writer is towards writing episodic narratives that are grouped together only in loose terms.

However, in Cervantes's great two-part masterpiece *Don Quixote* (1605, and 1615) the episodic dimension is controlled by the plot rather than *vice-versa*. The chief character is a retired gentleman Alonso Quixano, who is much given to reading tales of ancient chivalry—so much so, indeed, that he is no longer able to distinguish between reality and chivalric fiction. He adopts the noble name Don Quixote, and, in the company of his squire Sancho Panza seeks out adventures that will allow him to fulfil his chivalric dreams. However, the tragedy and comedy of his endeavour is that his dreams are realized only through delusional fantasy projections of chivalric meaning onto mundane situations and persons encountered in the course of travel. Indeed, the plot weaves into itself stories told by people whom Don Quixote encounters, and ideas about poetry and literature themselves. Quixote's tendency towards interference in situations that have nothing to do with him, and his failure to pay-off debts, mean that what happens in the narrative is a mixture of fantasy, and intrusions (often harsh and violent) from mundane reality.

In the second volume, Don Quixote and Panza have many other misadventures, until the Don Quixote decides to retire and live the humble life of a shepherd. At the end of the story, in the course of a mortal illness, Don Quixote regains his sanity and sees the misfortunes he has undergone (and caused for in others) in their true light. However, this directly engages one of the most extraordinary features of Cervantes novel. In the second volume, he weaves reflection about the narrative and its purpose (and about the authorship of his text) *into the narrative itself*, and has his central characters participate in this reflection. The effect is that the novel is simultaneously both a grand story *and* a questioning of the scope of fiction in its relation to reality.

This is the supreme paradox of Cervantes. *Don Quixote* is the work that establishes the idea of the novel as a thematic unity, and thus determines a format that sets a basic format that will be hugely influential. It also anticipates the questioning of this format, and the search for alternatives to it. In effect, it sets out the scope and limits of the novel as a storytelling art form.

III Theatre

Performance is basic to the development of dance, and theatre as modes of artistic expression. By definition, performance is determined by the relation between three things—a body engaged in vocal and/ or broader physical gestures, a physical setting where the performance occurs, and an audience to which it is directed.

Some such performances are focussed on the body itself and the performer's sense of it (as in dance). In these cases the performer may or may not be indifferent to his or her physical setting or to the question of an audience. The performance might focus on the body moving through space, or on the interpretation of movement through adopting a specific range of postures and actions regarded as meaningful in particular respects. It may involve occupation of, a specific physical or virtual setting(s). The body can physically occupy only one setting at any one time, but televisual or digital simulacra of it can be transmitted to other sites simultaneously, or with time lags.

The body and its physical setting in performance can be mediated by 'found' or prepared props, i.e., items which are not part of the body, nor a fixed part of the performance's physical setting, but which support its realization. Exceptionally, the 'found' element may include a physical part of the setting (imagine, for example, an outdoor performance of *Waiting for Godot*, which makes use of a withered tree growing at the site of the performance).

In practice, most forms of performance art will be directed towards an audience. Indeed, a *performance* (as opposed to a mere sequence of gratifying or expressive actions) is directed towards an audience (although there does not have to be one present—one can perform 'as if' for an audience). The performance can require some interventional response from the audience, or be indifferent to it, or set out only to engage its active attention for a certain period of time. The nature of the time involved is determined by the character of the performance.

A performance can be spontaneous (in response to another person, situation, or location) or it can be planned—as when a child does something which it knows will please an adult. If it follows a pre-given thematic—and especially if the thematic is based on any kind of script, then the work becomes a play. *A dance or play is a performance which can be repeated.*

The repetitions can follow a pre-worked out routine or a script, in some cases allowing for deviations from it, or, in other cases, improvisations around it. The routine or script can be memorized entirely or in part, or, in the case of the script, be read from on the spot.

Theatre gathers up, and as it were, formalizes aspects of performance and its settings. A theatre *per se* is a setting which has been designated or prepared for a performance. It functions so as to draw attention to what is performed, and to set it aside from the surrounding reality. This will mainly involve the setting being physically prepared (e.g., through a stage or other defined space, with the audience positioned around it or in front of it). More exceptionally (but perhaps more importantly for the *pre-history* of theatre), it may involve the designation of a natural site which embodies the aforementioned attention-directing features.

This separating structure of theatre does not have to be absolute. Indeed, it can serve as a vortex to absorb audience energies in order to change them and send them back into the world. All theatre—especially the politically charged work of a playwright such as Brecht—does this metaphorically, but some avant-garde tendencies might do it physically as well. In this case the theatre is a temporary setting in conceptual terms—although physically speaking, it might be fixed, and used again and again for this purpose.

If the term theatre is to be significant as an activity, then it must center on the presentation of plays. Completely spontaneous 'street theatre' is better described as performance to the degree that it does not involve repeatable material.

The foregoing descriptions, then, set out the very basic symbolic structure of performance art and its theatrical expression. Obviously, the features described cannot even begin to be linked in a strict chronological narrative. Rather various factors will emerge, be lost, re-emerge, and then establish themselves, over the course of time.

But there are two very important points. To reiterate; what we have described as 'performance' is *logically fundamental*—it is the basic symbolic function as embodied in this artistic idiom. Its basic features are rooted in behavioural possibilities which are phylogenetically and ontogenetically of the greatest importance.

Indeed, with more space, one might link 'performance' with 'mimesis' as a basic factor in the learning of language and social rules and in both identifying and distinguishing oneself from others. Theatre—with its self-conscious adaptation and development of basic 'performance' features—is, in effect, a higher-level model of all this.

As examples of significant theatrical achievement, we might consider two major figures from the ancient and modern worlds (respectively) Aeschylus, and Bertolt Brecht. Aeschylus is a foundational figure in the progressive articulation of theatrical drama as a medium. Before him, plays amounted mainly to oratorios accompanied by actions, or choral poetry accompanied by expressive dance. The chorus would exchange dialogue with a single actor switching from one character to others by using masks. Aeschylus's innovation of using a second actor transformed the performative scope of theatrical art. It increased the range of characters who could be represented (through mask-change) on stage, and allowed for more significant variations in the plot. Indeed, the use of a second actor meant that the chorus's role was diminished, and the plot became the major dramatic device.

Features such as this marked plays such as *Agamemnon*. However, it is in the only surviving Aeschylean trilogy *The Oresteia* (458 BC) that this enrichening of plot through enhanced performative scope becomes fully detectable. It is generally acknowledged that in Aeschylus, the motivations of the characters are not explored in any detail. Emphasis is placed instead upon the concrete decisions that they make. However, in the *Oresteia*, the introduction of the second actor allows for very considerable exploration of the grounds of the authority that govern decisions by mortals and God—issues concerning the relation between punishment and justice, and the nature of moral responsibility. Orestes revenged the murder of his father by killing his mother Clytemnestra, and for this it was ordained that he must be punished by the Furies. But, the nature of the crime committed by Clytemnestra means that Orestes has grounds to claim justification for his acts, and the trilogy explores his flight from the Furies, and the eventual grounds upon which an agreed principle of authority is established that can accommodate all parties. This received verdict on what is just is then assumed as a principle by Aeschylus's successor Sophocles. With the issue of authority resolved, Sophocles's introduction of a third actor allows him to address, in more detail, the motives that prompt individual characters to act in the way they do.

Aeschylus's simple innovation, then, in terms of the number of performers, enabled the progressive articulation of theatre. Of course, even more radical innovations are found in works by modern dramatists such as Bertolt Brecht. Brecht is useful for our present analysis because much of his work looks back to the model of classical drama. Progressive articulation often involves radical breaks, but these sometimes make us of important features of previous art. In Brecht's first play *Baal* (1918), for example, he employs heightened prose, songs, and even an introductory choral hymn. By restoring these classical features in a modern context, he gives them new meaning.

In the 1920s and 1930s Brecht wrote and directed many successful plays and was involved in devising scenarios for film, and in musicals—such as his collaboration with Kurt Weill on *The Threepenny Opera* (1928). From the mid 1920s, indeed, his creative work was often conducted in a deliberatively collectivist context including actors, designers, and other co-workers in dramatic production. By 1930, he had devised a principle of 'epic theatre' that attempted to overcome the opposition or rivalry between words, music, and production, by as it were, acknowledging their difference and allowing them to make their own unique contribution to the final work.

This overcoming of the division of media, was also accompanied in practice by radical changes to the performative basis of theatrical drama. Consider, for example, Brecht's great work *Mother Courage and Her Children* of 1939. The play covers the years from 1624 to 1636 inclusive. Mother Courage is an army camp-follower selling victuals, but rather than getting rich from the enterprise, over the course of the years, her children all die in the vicissitudes of war. However, she is not represented as a tragic heroine, but rather, as one who is a victim of war, and yet not altogether blameless for the things that befall her. The audience is distanced from empathy with her suffering towards reflection on the material circumstances that allowed it to happen and be sustained.

These effects are achieved through the relation between the plot itself and the use of things such as placards revealing the events of each scene, unusually intense or incongruous lighting, and changes of costume and character that actually take place on stage. Songs are also used throughout the play to emphasize points that the playwright is trying to make.

Brecht's 'epic' creation, then, uses some of the diverse means that were found in classical drama, but adds other features that break the imaginative distance between theatre and reality. In this way he achieves the kind of communal impact that was a feature of classical epic theatre, but he transforms the nature of that impact. Instead of the audience being imaginatively absorbed in the drama of the represented events and led to catharsis, its experience is agitated by intrusions from the real world. These make the drama politically telling by disrupting our comfortable expectations of the theatrical format, in ways that provoke thought about what the historical standardizations of that format have concealed.

IV Music

With the notable exception of Peter Kivy, there has been a strong tendency in aesthetics to link music to the expression of emotions.[9] However, what has not been much addressed in recent times is the particular way in which music is expressive. Simply saying it has expressive qualities is not enough—we need criteria of what makes expression in music unique.

Crowther has addressed this. In effect, he shows how music answers to the need for embodying narrative in an enduring and heightened sensuous form. His more immediate strategy is to go to the most basic origins of both the emotions and music. He argues, for example, that we can put ourselves in situations which we can be expected to cause a specific emotional response in us, but emotions themselves are not subject to the will. They are involuntary phenomena—that arise on the basis of our appraisals of people or states of affairs, or situations in terms of their positive or negative relation to our personal interests.

It is clear that whilst some symptoms of emotion (such as crying when grief-stricken) are natural, they can be modified by cultural circumstances and conditioning, leading sometimes to codified patterns of expressive behaviour. The voice is extremely significant in terms of this. Under normal circumstances it is subject to the will and has extremely wide range in terms of intonation and auditory volume. Whilst emotional states are involuntary, their vocal expression can incorporate voluntary gestures of the mouth and larynx that give rise to a controlled range of sounds. This allows for more controlled auditory expression of emotion. Language greatly extends the emotional inflections in the voice, allowing not only control and direction of the voice but also the augmentation of expression through the style of utterance. One can recognize the emotional tones of an utterance or conversation (e.g., angry voices) even when one cannot actually hear what is being said.

According to Crowther, music develops these features further. He notes with approval Susanne Langer's remark that:

> Music begins only when some formal factor—rhythm or melody—is recognized as a framework within which accent and intonation are elements in their own right, not chance attributes of individual speech.[10]

People can imitate the vocal gestures correlated with an emotion, so as to make some point in a conversation, or to deceive or amuse other people. In such cases, the vocal gesture is created with the intention of representing a possible emotional response rather than an actual one. Of course, it may be that sometimes such a vocal gesture is made to express an actual emotional state that the subject is in, but even here, the formalization—a song or whatever—means that we do not have to interpret it as the expression of some actual emotion which the vocalizer is 'in the grips of'. The expressed emotion is located in the domain of possibility, rather than that of actuality. Crowther uses, accordingly, the term virtual expression for emotion experienced in this context. There is a recognizable kind of emotion or range of emotions, but be we do not have to interpret them as states of the vocalizer. However, because they are vocal they express emotion at the same ontological level of some genuine manifestations of emotion (such as shouting angrily).

Crowther's point, then, is that in vocal music, we find a transition from a realm where real emotional responses are often articulated to emotions virtually expressed. Musical instruments allow this to be developed further, through making sound-pitch controllable, and thence subject to the will. Just as importantly, the use of instruments extends the virtual aspect of musical expression (i.e., the fact that what is expressed does not have to be an actual emotional state of an actual person). Of course, in order for an instrument to be played, a person is needed (under normal circumstances), but because we recognize that the sound is coming from the instrument, this means that it is quite obviously not—in literal terms, a directly occurrent emotional state of a person.

However, the sound so produced is still relevant to the ascription of such states, in more indirect terms. Because the sound has been recognizably produced through intentional human artifice—the playing of the instrument, and insofar as it is describable as 'sounding' tranquil, agitated or elated, etc., then we can rightly interpret it as being intended to communicate the possibility of a certain kind of emotionally significant auditory gesture. Instrumental music has its origin in vocal intonation but makes it more virtual through producing it from an instrument rather than from the body directly.

Crowther emphasizes that an important narrative dimension must be given consideration to. Individual musical units have their full significance only when heard in the context of the temporally developing whole of which they are a part. Each element is symbolically pregnant with what is to come, just as they are interpreted in terms of the parts which preceded their emergence. Crowther links this to narrative meaning, and directs us to the following remark by Schopenhauer.

> Melody is always a deviation from the keynote through a thousand crotchety wanderings up to the most painful discord. After this, it at last finds the keynote again, which expresses the satisfaction and composure of the will, but with which nothing more can be done, and the continuation of which would only be a wearisome and meaningless monotony.[11]

However fanciful Schopenhauer's seems by virtue of its exactness, Crowther insists that his basic insight is sound. He claims that:

> The tonal scale-system allows the relation between units of sound (be they vocal or instrumental) to be formalized to such a degree, that the narrative structure inherent in rhythm becomes much more complex. Themes, rhythms, and harmonies can develop and be described in the same sorts of terms which exemplify the emergence of emotional states in personal and group narratives.[12]

Indeed, we are told further that

> One might describe a piece as 'cheerful' or 'sad' but if there was to be no more to be said of it than that, it would not be a piece of much significance. The real substance of musical meaning lies in the way that one stage of recognizable emotional import is transformed into another-often in an extended way through the use of melodic, rhythmic, and harmonic factors that bear some analogy to the progressing tones of a conversation.[13]

This might be made more phenomenologically exact as follows. When a conversation develops, we note changes in how the voices of those involved change in terms of rhythm and intonation. This may range from the ins and outs of mere matter of fact tones, to a growing sense of urgency and developing import vis-à-vis what is said. There may also be a sense of matters have been resolved, and agreements reached. (Although Crowther does not remark on it, one can also see elements of this in actions—where people come together in a certain task or situation, engaging in actions that advance or retrogress matters). It is worth linking these points about to Cassirer, who notes that every meaningful conversation involves a kind of dialogical energy based on 'the twofolded direction of synthesis and analysis, separation and reintegration.'[14] He suggests, also, that this even extends to monologues with oneself. In fact, 'This double function of everything symbolic, the function of splitting and reunification, emerges still more clearly and convincing in *art*.'[15]

Crowther's interpretation of music links it to the various conversational phenomena just noted. He emphasizes that we can get a sense of the conversation, even if we cannot actually hear the content of what is being said. Just by listening to the vocal tones and rhythms of the conversation, we follow its cumulative narrative in terms of definite emotional inflections and 'conclusions'. Similarly, we may behold people engaged in physical actions and interactions *at a distance*, we may not know exactly what they are engaged in, but we can recognized when they are inaugurated and brought to a close, and key episodes in between.

The point is, according to Crowther, that these various dynamics of intonation and gesture—the discerned, as it were, *outlines* of narrative formation in conversation and action, are the features that basic symbolic function of music its very flesh. The musical meaning of the work as a whole has a developing texture of expressive properties that only cumulatively link to present specific emotional inflections. These motional characteristics are the growing matrices from which e passages of climax and/or resolution emerge. Virtual expression comprises both recognizable emotional gestures and the looser tonal narrative progressions from which they are cumulatively emergent.

As well as emotional associations, these progressions also provoke imaginings of possible action and situations (involving persons, events, and broader situations). These more existential associations are chiefly determined by the listener's personal experience of the listener, rather than the objectively accessible tonal character of the music. This objective guides our patterns of association, but the way these guidelines are realized reaches deep into as individuals—in much the same way as the transperceptual associational space of metaphor works.

Crowther emphasizes that virtual expression is not just some mere echo of our emotional life. We often have the impression that emotions are sudden dramatic things, but, of course, they are often much more sustained and complicated, things that develop rather than happen. And even when we are suddenly in the grips of an emotion, it is because our response is enabled by our appraisal of something in terms of its positive or negative significance for us. This means that it is intrinsically connected to the narrative of our life. However, this link between narrative and emotion is rarely noticed, because we are too much in the grips of the relevant emotion. In music, however, the narrative build-up to the great emotional outbursts become a source of pleasure in their own right, as well as in preparing us for the climax itself. Virtual expression elucidates and discloses the cognitive basis of emotion, making it aesthetically available.

Having linked music to virtual expression, Crowther gives consideration to the question of whether we have to make this link or not. Would it not be possible, for example, to analyse musical meaning on a purely formal basis—referring to the beauty of its melodies, harmonies, and rhythms? Crowther answers 'no'. He tells us that it is not just the beauty of the form, but its distinctiveness which is paramount. 'artistic originality is involved and this entails expressive factors insofar as a work's superiority to other created forms is, of necessity, a constitutive element in our appreciation. To enjoy form in this context is (at least in part) to be moved by, or to admire, the artist's treatment of it.'[16]

This conceptual linkage of formal and expressive factors is further developed by Crowther. He notes that one can refer to such things as 'violent' passages, and that even apparently 'pure' formal notions such as 'harmonious' 'clumsy' and 'elegant' entail some kind of expressive association, through being derived, analogically, from broader aspects of human interaction and gesture. Even with such an apparently neutral term such as 'balance', sounds can be described literally as 'balanced' only in a quantitative sense (e.g., through their loudness or duration). Qualitative 'formal' notions of balance apply only through analogy with relationships of gesture and /or action. Things like fugues, first subjects, recapitulations, sonatas, and such things, embody formal structures. But to appreciate these forms requires more than recognition of their structure. We are concerned with what the forms do, rather than what they are. And this means that the narrative features that Crowther links to virtual expression must be involved, no matter how subtle this involvement may be.

On the basis of a Cassirean approach we might see the formalist emphasis in music appreciation as a relatively modern phenomenon, arising through the progressive articulation of the more basic idiom of virtual expression. As the basic symbolic function of musical meaning it remains present even when its emphases push towards something more abstract. As Crowther observes,

> If one is highly cultivated in the understanding of musical form, one may be overwhelmed by formal issues—but only because such forms have already been established as having broader existential import. Through our broader culture they are transmitted to us as representations of virtual expression. And even if our formal appreciation does not take this into account explicitly in the understanding of musical form, may be overwhelmed by formal issues—but only because such forms have already been established as having broader existential import. Through our broader culture they are transmitted to us as representations of virtual expression. . . . If a formalist rejects this account, it is incumbent

> on him or her to offer an alternative which can explain the genesis and transcultural ubiquity of music.[17]

It is interesting that when we experience the visual arts, literature, and the performing arts, we often achieve some aesthetic empathy with the specific individuals who are being represented, and the way that the artist articulates the narrative concerning them. However, Crowther notes that this involves a constraint. The figures who we are engaging with are other than us—definite individuals who have notional histories and destinies other than ours. This means that our identification with them (no matter how imaginatively engaged we may become) is ultimately distinguished from ours.

Instrumental music is less constrained. It characterizes particular kinds of emotions and the quasi-narratives from which they are emergent, but these features are not tied to any recognizable individual bearer of them. The auditory gestures that the composer embodies in the work may be, in some sense, about' his or her experience, or even that of a particular imaginary person, or even of a social group. However, the work is aesthetically meaningful without it having to be referred to any specific individual's experience. Music as virtual expression, presents a form of experiential possibility without any definitive owner. We may identify with the composer or performer's creative way of presenting the narrative progressions and emotional foci—with his or her style of realizing musical form—but the progression of forms so realized is not tied to a specific individual or individuals in the way that characters in literature or people or places represented in pictures are.

This means that the division between music and those who compose it, perform it, or listen to it, is diminished. The gestural and narrative richness of the work is not tied to any definite individual (for the reasons noted earlier) so we can enjoy it, perhaps even live it on our own terms in a way that is not possible with other art media. Of course, there can be stories or poems where the identity of the subject or narrator, or times and places involved, are kept anonymous. However, whereas in literature this is an unusual particular *use* of fundamental symbolic structures, in music it is intrinsic to the basic symbolic function itself. Music involves expression which, by its nature, is not assigned to identified individuals. As Crowther puts it,

> This . . . creates a sense that we are in the music rather than merely encountering it as an 'external' object of auditory experience. And whilst any artwork allows empathic identification with its producer . . . the lack of individual reference in the musical work gives this identification a phenomenological closeness and intimacy which is not available from other media.[18]

Crowther himself discusses a wide number of composers in the course of his analysis. He does not explicitly describe them as progressively articulating the medium, but it is clear that this is his meaning. In discussing the

Prelude to Wagner's *Tristan and Isolde*, for example, he suggests that general stylistic traits of Wagner's developing style (and perhaps late-Romanticism itself) are given a particular chromatic emphasis with subtleties that not only extend the musical scope of the tonal system, but also point towards atonal possibilities. In Crowther's words, 'This is more than the achievement of an individual and influential style. It is one which changes the terms in which musical creativity is itself conceived. And through that, it offers new ways of addressing virtual expression. It not only expands musical tradition, it opens up a new direction for it.'[19]

However, the most interesting example of progressive articulation in music that Crowther discusses is that of the living composer Steve Reich. His 1965 work *It's Gonna Rain* uses the 'phase' technique (where multiple tapes are played simultaneously, slightly out of synch with each other, and recorded) as basic compositional structure. In it, the musical meaning is created from repetitions of sound gradually varied so as to transform and merge almost imperceptibly into new structures. Here, techniques originally used for the mechanical reproduction of sound are reconfigured as a basis for composition.

Subsequent works from the 1970s such as *Six Pianos*, and *Variations for Strings and Wind Instruments* use the repetitions of the phase technique as an inspiration for compositions involving more traditional instruments and ensembles. Tapes and phasing effects themselves are not actually used. Instead, Reich's compositions evoke the rhythmic drives, and precise layered repetitions that the mechanical reproduction of sound on tape through phasing draws attention to. In this way, Reich extends the scope of tonal music through highly unorthodox means.

Reich's masterpiece *Different Trains* of 1989 brings all the material explored since 1965 into combination. Compositionally speaking, the work's dynamic is one of insistent rhythmic repetition and variation, with audio tapes providing exact elements of vocal narrative, as well as contributing to the developing rhythmic drive.

The work based on two parallel but incongruent dimensions of Jewish experience from the Second World War. One is that of Reich's own childhood rail travels across America through the travails of his parents' divorce, and the other is that the transportation of Holocaust victims by rail in Europe. Through historical and geographical circumstances Reich was spared the horrendous destinies of his European counterparts. The clash of chance and grim inevitability at issue in this drives the work forward through extraordinary combinations of reminiscence, horror, and redemption.

Crowther suggests that *Different Trains* is the most important statement of the 'minimalist' tendency in postmodern music—one that brings its expressive potential to fulfilment. He does not, however, consider the question of this example being at odds with some of the general points about music noted earlier. Reich's piece, for example gets its power from being historically and personally highly specific in its thematic material. We cannot simply slip into and inhabit this music.

However, to be fair, Crowther does seem to base his analysis mainly on the case of purely instrumental music. This means that his position requires further work. He might argue that—as in Brecht's epic theatre—music and song, and the written word mutually illuminate one another through their proximity within a specific work, rather than on the basis of some qualitative fusion. But, there again, it may be possible for Crowther to make a case for a more unified theory of music and vocals through thinking this relation through in terms of his explanation of the origins of music in the vocal gesture.

Notes

1 Crowther has published little on literature or theatre as art forms. Our account of them in this chapter is based on a development of detailed lecture notes taken by Kopsiafti in classes taught by Crowther in the 1990s; and from the discussion of literary metaphor in Chapter 6 of *DACC*. In contrast Crowther has published an extremely long dedicated chapter on musical meaning as Chapter 7 of *DACC*, and it is on this that our account of his theory is closely based.
2 See <https://en.wikipedia.org/wiki/We_shall_fight_on_the_beaches>. Downloaded 12/07/2014.
3 Paul Ricoeur, *The Rule of Metaphor*, trans. Robert Czerny, John Costello, Katherine McLaughlin (London: Routledge, 1994), 65.
4 Nelson Goodman, *Languages of Art* (Indianapolis: Hackett, 1976), 69.
5 As we have seen, Crowther uses the transperceptual mainly as a basis for formulating a general theory of meaning for abstract art. (See, for example, Chapter 9 of *TPMA*.) Our point is that the concept has a more general application in aesthetics.
6 There are some clear affinities between transperceptual space and what Merleau-Ponty calls the 'invisible' or 'Flesh'. Crowther has noted the affinity, but has recently offered a very critical account of the shortcomings of Merleau-Ponty's position, in terms of his lack of attentiveness to the full implications of the role of reflective knowledge (an issue we considered earlier). See *TPMA*, Chapter 4.
7 We have translated Mallarme literally. The nearest version to our approach is *Mallarme*, ed. and trans. by Anthony Hartley (Harmondsworth: Penguin, 1970), xviii.
8 For a published (near equivalent) translation see Hartley's *Mallarme*, Ibid., p. 63.
9 See, for example, Kivy's *New Essays on Musical Understanding* (Oxford: Clarendon Press, 2001).
10 Susanne K. Langer, *Feeling and Form: A Theory of Art* (London: Routledge and Kegan-Paul, 1953), 142.
11 Arthur Schopenhauer, *The World as Will and Representation*, trans. E.J. Payne (New York: Dover Publications, 1980), I, p. 321.
12 Crowther, *DACC*, p. 178.
13 Ibid.
14 Cassirer, *LCS*, p. 54.
15 Ibid., pp. 154–155.
16 Crowther, *DACC*, p. 180.
17 Ibid., p. 182.
18 Ibid., p. 189.
19 Ibid., pp. 198–19.

Conclusion
Aesthetic Cognition and the Future

In this book we have hoped to show how aesthetics as a discipline might be revitalized by addressing Cassirer's account of aesthetic cognition and Crowther's approach to it. Cassirer's general philosophy is of enduring worth, and his idea of the basic symbolic function offers a structure that is flexible to unify our conceptual framework and accommodate the diversity of its components. Although Cassirer described himself as a critical idealist, we have shown how his work is better described as a corporeal phenomenology, and that this gives his work affinity with Merleau-Ponty.

However, Cassirer emphasizes the importance of reflective knowledge and does justice to it in a way that Merleau-Ponty does not. Indeed, the whole point the philosophy of symbolic form is to connect how act in the world at a basic level, with how we think about in more general terms. He rightly emphasizes that the reflective standpoint distances from the world in order to return us to it.

This epistemology has great strength, but we noted that Cassirer did not have a properly developed theory of imagination. He needs this both to account for the relation between objective knowledge and knowledge of self, and because of the importance he assigns to imagination in his aesthetics. We argued that Crowther's theory is the perfect complement to this. It sets out an approach that assigns a transcendental role to imagination without being tied to the technicalities of Kant's idealism, and it offers an ontogeny and phenomenological account that reconfigures how we must think about the unity of the self. It shows how imagination's creative stylizing features have an importance in making past, future, and counterfactuality matter to us. Indeed, the blending between imagination and its object explains the lingering positive role of mythic thought in all consciousness. Crowther also offers a distinctive interpretation of the self as an aesthetic narrative structure. Expanding on Crowther's approach, we emphasized the frailty and transience of the conscious states and imaginings that constitute the *act-association continuum* in which all our practical activity takes place.

This material allows us to illuminate the further scope of the basic symbolic function. Indeed, Crowther's account of imagination and the narrative unity of the self allowed us to identify those needs which the creation of art answers, and the motives—over and above survival considerations—that the human subject has for developing and progressively articulating symbolic form.

Whilst the self has always had this structure, it has not been theorized in explicit terms. This why the development of aesthetics as a discipline is important. It is not only the diversification of style in the arts that progressively articulates aesthetic cognition; the rise of aesthetics *itself* is implicated in this development. The *concept* of art and the aesthetic as a distinct form of knowledge requires the proper kind of theoretical reflection as well as new forms of art practice. This is why we devoted so much attention to Cassirer's mapping of Enlightenment aesthetics. In that era, changing conceptions of the relation between the artwork and imagination, and the objective and subjective aspects of experience, and the importance of the questions of disinterestedness and universal validity, were all raised and pursued in different ways. These are the concepts that point towards the enduring importance of aesthetic cognition.

It is precisely this cognitivist dimension which contemporary Anglo-American aesthetics has lost sight of. Through this, it has become marginalized in terms of relevance to the contemporary world of art practice and appreciation. The Continental tradition has considered only rather narrow aspects of aesthetic cognition, or more general points about the cultural significance of art, and the conditions of its consumption.

The only way of recovering the situation and creating a vital future for aesthetics is to focus on how aesthetic cognition originates and becomes theoretically and artistically embodied and then progressively articulated. Chapters 1 and 2 explained the origins, Chapters 3 and 4 the theoretical embodiment, and Chapters 5 and 6 considered the artistic embodiments and their progressive articulation. Instead of looking at talk about art, we considered how the creation of works in individual media changes the world—each in its own unique way. And instead of mentioning a few artists along the way, we followed Crowther's cues in taking the trouble to look in some detail at the achievements of particular artists whose individual development progressively articulates their chosen medium.

There are some important lessons for the future of aesthetics that can be learned from this approach; some of them are particular, and others of much more general importance. In terms of particular considerations, the following come to mind. Cassirer's thought, in particular its analyses of language, and of myth, offer a potential goldmine of insights especially for literary aesthetics. We showed an important overlap between imagination as such and mythic thinking. This can be taken much further than we have been able to do. It may, indeed, have unsuspected ramifications.

From Crowther, as already noted, we draw the moral that the main focus of attention in the aesthetics of art, is on how the artist's style of making changes our perception and conception of the world and our relation to it. To bring this out involves great emphasis on what particular artists do and what this, in turn, implies for the medium in question. We have shown Crowther's approach to these issues, but there are surely many ways of doing this besides his strategies. Likewise in his thinking about the aesthetics of the self. His approach is very distinctive—we think especially here of the emphasis he places on the momentary experience as a focus of consciousness, and on its aesthetic role. But again there are many other ways in which narrative structure in the self and the way it is engaged by particular modes and/or individual works of art.

There is another fruitful particular possibility raised by Crowther's approach—maybe the most intriguing of all. In this book we have been able only to touch on the question of aesthetic experiences of nature as part of our more general problematic. This is a topic of the greatest interest in itself. A cognitivist approach wants to argue that the aesthetic configuration changes our perception of world and self. It is surely the case that this can happen through the experience of nature as well as through art. We can reasonably assume that this might involve a fascinating range of features—engaging with metaphysical and religious insights as well as the narrative of the self. However, it might be that the aesthetics of nature actually marks out the limits of an aesthetic cognitivist approach, and is resistant to its aspirations. (The present writers would love to explore these issues themselves, but the practical circumstances of our present institutional commitments most likely preclude it.)

Over and above these individual possibilities, the Cassirer-Crowther approach points towards two important general developments, one concerning the definition of art, and the other concerning more general points.

First, the problem with the definition of art concerns the way it has been usurped by specifically western and visual-orientated approaches. Duchamp's ready-mades suggest that art can be anything that the artist chooses to regard as art. The important point is not making the work of art, but—to use Crowther's terms—an act of designation. Now, as we saw earlier, Crowther does not rule out ready-mades being works of art, but he does see them as secondary and extremely derivative. The central core of art involves making and or designing in terms of a sensuous or imaginatively intended medium. Interestingly, in fields such as film, literature, theatre, and music, peripheral activities that might have parallels with Duchamp's work (such as some of John Cage's music) have been influential but have by no means been invested with the kind of authority that Duchamp's ready-mades have for visual art. In the non-visual arts, in other words, a primacy remains on complex processes of design and/or making where the cognitive dimension remains the main focus. And, even in the visual arts, in non-western cultures, the great bulk of artistic creation (that follows indigenous rather than western idioms) tends to emphasize processes of making that change how the world appears through being represented. Indeed, these aesthetic configurations are often used for quite deep ritual functions concerning the individual human's relation to nature, ancestors, God, and other relevant ideas.

It might seem that this makes such artefacts into things of only ritual significance. However, Crowther rightly asks,

> how is it possible for non-western representation to be invested with . . . extraordinary levels of 'ritual' metaphysical and religious significance . . .? What is it about the making of such representations which facilitates the belief that they can bond the creative process, the work, the world, and the audience in an intimate living relationship?[1]

Crowther notes also that some cultures take sensuous representation as to actually be alive or give control over what is represented, and asks again what explains these beliefs. Why do people from so many different parts of the earth and different cultures feel so confident about the idea of mimetic artefacts having this sort of power?

We might suppose that it is the sensory or imaginative vividness of mimesis that give it more immediate cognitive impact than linguistic reports. Such mimesis has a sensory vividness that is attractive at a biological level, and this explains why people would want to use it in rituals and the like. There is no need to make reference to such concepts as art and the aesthetic.

However, Crowther also suggests that such simple sensory attractiveness is not enough. We must accept as well that there is something about making the likeness, *as such*, which is intrinsically satisfying. One is literally changing how the world appears and it is this pleasure in *aesthetic* formation which is paramount. As Crowther puts it

'These intrinsically valuable experiences facilitate the belief that representation is the kind of privileged activity which can realize . . . metaphysical and religious effects . . . Through the power of aesthetic embodiment, these abstract ideas find an appropriate and distinctive sensible manifestation. Belief in their ritual potency is thus *enabled* by their aesthetic power.'[2]

Hence whilst not all cultures have concepts of art and the aesthetic, artistic and aesthetic meaning are at the source of the ritual investment that they make in the practical efficacy of the relevant artefacts.

Aesthetic cognition, then, can establish the definition of art on a genuinely universal basis instead, of defining it in a way that primarily reflects western interests and issues in the visual arts. This outcome should not be found surprising—because the formative power of the aesthetic and artistic has its origins in imagination and the unity of the self (as we showed in Chapter 2) and these features. In turn, are distinctive embodiments of the basic symbolic principle.

Cassirer-Crowther aesthetic cognitivism, then, establishes that the aesthetic and artistic are not luxurious commodities, but express and develop features basic to our humanity. This creates the possibility of a genuine Copernican turn in aesthetics where the emphasis would switch from the conditions under which artworks are engaged with from an audience viewpoint, to the ontology of how such works—through their intrinsic features and relation to other works in the same medium—transform the appearance of reality. It is this transformation that is the foundation of aesthetic cognition. If aesthetics can follow the Cassire-Crowther approach, issues might arise that would engage the interest of both artists and a broader public.

There would be other potential benefits. The present writers do not have experience of philosophy of education, but it seems likely that for those involved in art education at every level, Cassirer-Crowther aesthetic cognitivism is a potential goldmine of ideas for explaining and justifying education in the arts.

There is a bigger implication also. To be human is develop one's powers of reason, sensuous receptivity, and emotional life, especially in terms of empathy. The development of symbolic forms and the understanding of their progressive articulation, extend the scope of our humanity. Initiation into a proper range of them offers the cognitive means to *critical autonomy*. This means that an agent has the capacity to recognize how the world is, in the most general sense, and to make sense of the particular situations it presents. He or she is also to posit and choose between different possible strategies of responsive or anticipatory behaviour, and to evaluate different ways of giving meaning to existence in general. Symbolic forms are not simply things one learns. Through being initiated into them, one become an agent in dynamic terms, instead of one who just accepts the status quo.

On these terms, then, as well as its own intrinsic excellences, Cassirer-Crowther aesthetic cognitivism fits into a bigger picture—that of clarifying the importance of the aesthetic domain's contribution to the creation of critically autonomous agents, as such. Aesthetics has enduring and expanding relevance for self-development.

Indeed, considered together, Cassirer's and Crowther's work forms a whole that is far more than the two components. We have explored only some of the more salient opportunities for linking them in the context of aesthetics. There are many more—in epistemology, ethics, and, in particular, the philosophy of culture. Given the amount of interest in the theory of culture across so many academic disciplines, it is surprising that philosophers have not made it more central as an area of study. One reason for this is a lack of systematic philosophical approaches that could provide a point of orientation. However, in twentieth-century philosophy, the Croce-Collingwood pairing picked out a philosophical affinity that created important possibilities for the philosophy of history and culture as well as for aesthetics. We hope that the Cassirer-Crowther link will likewise create a powerful intellectual resource in the present century for advancing philosophy of culture as a major area of research.

Notes

1 *DACC*, p. 51.
2 Ibid., p. 52.

Bibliography

Barash, Jeffrey A., ed. *The Symbolic Construction of Reality: The Legacy of Ernst Cassirer*. Chicago and London: University of Chicago Press, 2008.

Bayer, Thora I. *Cassirer's Metaphysics of Symbolic Forms*. London and New Haven: Yale University Press, 2001.

Bayer, Thora I. "Art as Symbolic Form: Cassirer on the Educational Value of Art". *The Journal of Aesthetic Education* 40, no. 4 (Winter 2006): 51–64.

Bergson, Henri. *Time and Freewill*. Trans. F. Pogson. London: George Allen and Unwin, 1959.

Bundegaard, Peer F. "The Grammar of Aesthetic Intuition: On Ernst Cassirer's Philosophy of Symbolic Form in the Visual Arts". *Synthese* 179, no. 1 (2011): 3–17.

Burke, Edmund. *A Philosophical Enquiry into the Origin of Our Ideas of the Sublime and Beautiful*. Oxford: Oxford University Press, 2008.

Casey, Edward S. *Imagining: A Phenomenological Study*. 2nd Edition. Bloomington: Indiana University Press, 2000.

Deleuze, Gilles. *Francis Bacon: The Logic of Sensation*. Minneapolis: University of Minnesota Press, 2005.

Dickie, George. *The Century of Taste: The Philosophical Odyssey of Taste in the Eighteenth Century*. Oxford: Oxford University Press, 1996.

Evans, Gareth. *Varieties of Reference*. Oxford: Clarendon Press, 1982.

Gadamer, Hans-Georg. *Truth and Method*. Trans. William Glen-Doepel. London: Sheed and Ward, 1979.

Goethe, Johann W. von. *Goethe on Art*. Ed. and trans. J. Gage. Berkeley and Los Angeles: University of California Press, 1980.

Goodman, Nelson. *Languages of Art*. Indianapolis: Hackett, 1976.

Heidegger, Martin. *Basic Writings*. 1st Harper Perennial Modern Thought Edition. Ed. David F. Krell. New York: Harper Collins, 2008.

Johnson, Mark. *The Body in the Mind: The Bodily Basis of Meaning and Imagination*. London: University of Chicago Press, 1990.

Kant, Immanuel. *The Critique of Judgment*. Trans. J.C. Meredith. Oxford: Oxford University Press, 1973.

Kant, Immanuel. *The Critique of Pure Reason*. Trans. N. Kemp-Smith. London: Macmillan, 1989.

Kieran, Matthew. *Revealing Art: Why Art Matters*. London and New York: Routledge, 2004.

Kivy, Peter. *New Essays on Musical Understanding*. Oxford: Clarendon Press, 2001.

Krois, John M. *Cassirer: Symbolic Forms and History*. London and New Haven: Yale University Press, 1987.

Lamarque, Peter and Stein H. Olsen. *Truth, Fiction, and Literature: A Philosophical Perspective*. Oxford: Oxford University Press, 1996.

Langer, Susanne K. *Feeling and Form: A Theory of Art*. London: Routledge and Kegan-Paul, 1953.

Laushke, Marion. *Ästhetik im Zeichen des Menschen: Die ästhetische Vorgeschichte der Symbolphilosophie Ernst Cassirers und die symbolische Form der Kunst*. Hamburg: Felix Meiner, 2007.

Lechte, John. *Genealogy and Ontology of the Western Image and Its Digital Future*. London and New York: Routledge, 2012.

Livingston, Paisley. *Art and Intention: A Philosophical Study*. Oxford: Clarendon Press, 2007.

Lofts, S. G. *Ernst Cassirer: A "Repetition" of Modernity*. Albany, NY: State University of New York Press, 2000.

Lopes, Dominic M. *Understanding Pictures*. Oxford: Oxford University Press, 1996.

Lyotard, Jean-Francois. *The Inhuman*. Trans. Geoff Bennington and Rachel Bowlby. Cambridge: Polity Press, 1991.

MacIntyre, Alasdair. *After Virtue: A Study in Moral Theory*. London: Duckworth, 1985.

Mallarmé, Stéphane. *Mallarmé*. Ed. and trans. by Anthony Hartley. Harmondsworth: Penguin, 1970.

McGinn, Colin. *Mindsight*. Harvard: Harvard University Press, 2004.

Merleau-Ponty, Maurice. *The Primacy of Perception*. Ed. James E. Edie. Trans. Carleton Dallery. Evanston, IL: Northwestern University Press, 1964.

Meskin, Aaron. "Aesthetic Testimony: What Can We Learn from Others About Beauty and Art?" *Philosophy and Phenomenological Research* 69, no. 1 (2004): 65–91.

Metta, Carmen. *Forma e figura. Sul problema della rappresentazione in Ernst Cassirer e Paul Klee*. Macerata: Quodlibet, 2009.

Nancy, Jean-Luc. *The Ground of the Image*. Trans. Jeff Fort. New York: Fordham University Press, 2005.

Nietzsche, Friedrich. "The Birth of Tragedy". In *The Philosophy of Nietzsche*. New York: The Modern Library, 1954, pp. 947–1088.

Paetzold, Heinz. *The Symbolic Language of Culture, Fine Arts and Architecture: Consequences of Cassirer and Goodman*. Trondheim: FF Edition, 1997.

Pillemer, David. *Momentous Events, Vivid Memories*. Cambridge, MA: Harvard University Press, 1998.

Ranciere, Jacques. *The Future of the Image*. Trans. Gregory Elliott. London and New York: Verso, 2007.

Ricoeur, Paul. *The Rule of Metaphor*. Trans. Robert Czerny, John Costello and Katherine McLaughlin. London: Routledge, 1994.

Santayana, George. *The Sense of Beauty*. New York: Scribner and Sons, 1896.

Sartre, Jean-Paul. *The Imaginary: A Phenomenological Psychology of the Imagination*. Trans. Jonathan Webber. London: Routledge, 2004.

Schiller, Friedrich. *Letters on the Aesthetic Education of Mankind*. Trans. E. Wilkinson and Willoughby. Oxford: Oxford University Press, 1984.

Schopenhauer, Arthur. *The World as Will and Representation*. Vol. I. Trans. E.J. Payne. New York: Dover Publications, 1980.

Skidelsky, Robert. *Ernst Cassirer: The Last Philosopher of Culture*. Princeton: Princeton University Press, 2008.

Verene, Donald P. "Cassirer on Culture". *International Philosophical Quarterly* 22 (1982): 133–144.

Verene, Donald P. *The Origins of the Philosophy of Symbolic Forms: Kant, Hegel, and Cassirer*. Evanston, IL: Northwestern University Press, 2011.

Wollheim, Richard. *Painting as an Art*. London and New York: Thames and Hudson, 1987.

Zangwill, Nick. *The Metaphysics of Beauty*. Ithaca: Cornell University Press, 2001.

Index

For Product Safety Concerns and Information please contact our EU representative GPSR@taylorandfrancis.com Taylor & Francis Verlag GmbH, Kaufingerstraße 24, 80331 München, Germany